NAVAL ARCHITECTURE
FOR NON-NAVAL ARCHITECTS

Lester Rosenblatt's ketch *Rosa II.* Courtesy M. Rosenblatt and Son, Inc.

NAVAL ARCHITECTURE FOR NON-NAVAL ARCHITECTS

by

Harry Benford

Published by
The Society of Naval Architects and Marine Engineers
601 Pavonia Avenue
Jersey City, NJ

ISBN 0-939773-08-2

To Bill Zimmie
entrepreneur, engineer,
loyal friend

PREFACE

This book was written at the behest of the Education Committee of the Society of Naval Architects and Marine Engineers. Members of that committee, under the chairmanship of the late William E. Zimmie, perceived its need. They were aware of the large numbers of people whose work or leisure activities bring them into contact with the maritime world. Such people have no need for a deep understanding of naval architecture, but they can benefit from a sound introduction to the subject.

This book, then, is written primarily for the benefit of professionals such as oceanographers and other scientists and engineers who need to know something of the technology of floating craft. Interested readers may include naval architects' spouses, while others may be found among yachtsmen or anyone embarking on a shipboard cruise. Still others may include marine underwriters, admiralty lawyers, or high school students contemplating a career in the marine field. Finally, of course, there may be a wide readership among individuals whose intellectual curiosity may be piqued by the whys and wherefores of ships and boats and by the fascinations of the sea.

If floating craft play any sort of role in your work or hobby, you are well advised to teach yourself something of their capabilities and limitations. Perhaps you will see the need to communicate with naval architects. This book is intended to help. From it you will learn enough of the lingo and basic technology to exchange thoughts with members of that profession.

Practicing naval architects may be involved in a wide range of activities: creative design, analysis, management, production, fleet operations, and so forth. In the strictest sense naval architects create designs for all manner of floating craft (including their machinery and components). Their engineering work is a mixture of art and science. Proposing the initial, conceptual design is largely art; analyzing and refining it applies techniques based on sound scientific principles. In this book I concentrate on those two aspects of naval architecture: design and analysis. Moreover, because art is largely intuitive and therefore hard to explain, most of what I have to tell you will concern analysis. This will involve answering questions about such matters as the proposed vessel's seaworthiness, structural integrity, powering requirements, and functional capability.

You will find that my coverage focuses on conventional ships or boats of whatever purpose and refers only lightly, if at all, to offshore platforms, hydrofoil craft, air-cushion vehicles, and other nonconventional configurations.

The technical discussions will be almost entirely qualitative rather than quantitative. Most or all of what I have to say will be easily understood by anyone with a high school knowledge of physics and mathematics. Appended to each chapter will be found a list of recommended references for further independent study; a complete Glossary defines possibly unfamiliar terms.

Good luck to you as you now embark on this introduction to the beguiling mysteries of naval architecture.

ACKNOWLEDGMENTS

Writing a book of this nature is not easy. One problem is that of assessing the perspective of the intended reader. For help in this, I turned to three non-naval architects who were kind enough to read and criticize preliminary drafts. If the book is now readily understandable by its intended readers, those three reviewers' questions and comments deserve the credit. Let me thank them by name: Carolyn Churchill, Lorelle Meadows, and Russell Moll. To these names I must add that of my wife, Betty, who also served as proofreader of the semifinal draft.

A second problem I faced was the danger of being unaware of some of the current conventions among practicing naval architects. To overcome that possibility, I enlisted the help of many active members of the profession. Each was asked to criticize one or two chapters, and each responded with enthusiasm. (Some former students enjoyed particular pleasure in taking me to task, not only for technical matters, but for a few mistakes in grammar.) I am grateful for all their expert assistance and hereby register my indebtedness to Noel Bassett, Jay Benford, John Boylston, Joseph Fischer, Peter Fisher, Bruce Foss, Timothy Graul, William Guild, and Roy Harrington.

The list continues with William Hunley, Miklos Kossa, Barbara Lamb, Naresh Maniar, Richards Miller, George Plude, Donald Roseman, and Richard Suehrstedt.

Three of my colleagues showed considerable patience in advising me on matters within their own areas of expertise. They were Robert Beck, William Vorus, and John Woodward.

I am also grateful for miscellaneous forms of assistance from Francis Cagliari, John Daidola, John Flipse, Meade Gougeon, Ken Lo, Thomas Mackey, James Nivin, Lester Rosenblatt, Gregory Stewart, Charles Wilson, and Eileen Zimmie.

All members of our secretarial staff deserve credit for their frequent rescue missions on my behalf. I want to thank them for their ever-cheerful help: Mary Gibbons, Virginia Konz, Luella Miller, Lisa Payton, and Karen Trevino.

The original drawings were prepared by Rodney Hill. I appreciate the patience he showed as I subjected him to repeated requests for changes.

Finally, I want to register my thanks to Michael Parsons, who served as my contact person with the Education Committee of the Society of Naval Architects and Marine Engineers. His support and encouragement are genuinely appreciated.

Symbols and Abbreviations

A_x maximum cross-sectional area of the underwater hull
$A_{\otimes}$ cross-sectional area of the underwater hull at mid-length
a small area
AAC average annual cost
AC alternating current
AP after perpendicular
Asst assistant
A_{wp} area of a waterplane
B beam, also location of the center of buoyancy
B' shifted location of the center of buoyancy
$B/2$ half beam
BHP brake horsepower
BM metacentric radius
Btu British thermal units
Cab cabinet
C_{B} block coefficient
C_x maximum section coefficient
$C_{\otimes}$ midship coefficient
cm centimeter
CN cubic number
C_{p} prismatic coefficient
C_{wp} waterplane coefficient
CP controllable pitch
D depth (of hull), also propeller diameter
d any distance
DC direct current
DCF discounted cash flow rate of return
Dept department
Dn down
DWL designed waterline
EHP effective horsepower
engr engineer or engineering
F degrees Fahrenheit
FN Froude number
FP forward perpendicular
ft feet
G center of gravity
g acceleration owing to gravity
GM metacentric height
GRP glass-reinforced plastic
GZ horizontal separation between buoyancy and weight, that is, the righting arm
I moment of inertia

I-beam a structural member with cross-sectional shape somewhat like the capital letter I
I/y section modulus
K location of the keel, that is, the baseline; also the ship's transverse radius of gyration
KB height of the center of buoyancy above the baseline, that is, VCB
kg kilograms
kJ kilojoules
KM height of the metacenter above the baseline
KP kingpost
L any length
Lav lavatory, that is, washbowl
LBP length between perpendiculars
LCB longitudinal center of buoyancy
LCC life-cycle cost
LCF longitudinal center of flotation
LOA length overall
LWL length on waterline
M metacenter, also bending moment
m meters
MCT moment to change trim
M_L longitudinal metacenter
NA neutral axis
NPV net present value
P pitch of propeller
psi pounds per square inch
r radius
Recr recreation
R_f full-scale frictional resistance
r_f model frictional resistance
R_r full-scale residuary resistance
r_r model residuary resistance
R_t full-scale total resistance
r_t model total resistance
RFR required freight rate
RM righting moment
Rm room
S station spacing, also section modulus
s stress
SHP shaft horsepower
SI Systeme International d'Unites
sin sine
SM Simpson's multiplier(s)
Sta station
T draft, also the time of one complete roll (period)

tan	tangent
T&S	toilet and shower
V	volume of displacement
V_k	speed, knots
v	speed, feet per second
$\frac{V_k}{\sqrt{L}}$	speed-length ratio (length in feet)
VCB	vertical center of buoyancy
W	any weight
WC	water closet, that is, toilet
WS	wetted surface
y	any offset

Symbols and Greek letters

ℬ	baseline
℄	centerline
Δ	displacement
$\frac{\Delta}{\left(\frac{L}{100}\right)^3}$	displacement-length ratio
⊗	amidships, that is, mid-length
⊗$\frac{L}{2}$	midship half-length
θ	angle of heel
$	welded butt joint
∠	angle iron

CONTENTS

CHAPTER I

LET'S GET UNDER WAY

But first, do you have the bad habit of skipping the preface in books? (I do.) If you have done so here, please go back and read what I have to say. ***It really is important.***

1. What Floating Craft Do

All the boats and ships in the world can be placed into five categories according to their primary function. In no particular order, these are

a. To transport something: cargo, passengers, or whatever.
b. To provide recreation.
c. To provide military capability.
d. To provide services to other craft.
e. To provide a platform for scientific research or other miscellaneous functions.

We find this sort of breakdown useful because the economic, political, and human considerations affecting design differ greatly in emphasis between them.

2. Basic Requirements

We have already said that naval architecture is a mixture of art and science. Much of this is brought about because any sort of floating craft must be designed to meet many conflicting requirements. Considerable subjective judgment (art) is needed to effect successful compromises between these conflicting needs and to blend them into a harmonious whole.

First of all, of course, a floating craft must be buoyant. It must be buildable and maintainable. It must to one degree or another be economic both to build and to operate. It should also to one degree or another be pleasing to the eye. Finally, and of primary importance, it must embody seaworthiness appropriate to its area of operation and to the degree of expertise of its intended crew.

What is involved in seaworthiness? Here again we have a catalog of conflicting demands. A seaworthy ship is not about to capsize; that is, it must be stable. That encourages great width (what we call beam). On the other hand, a seaworthy (and economic) ship must be easily propelled, and that encourages a narrow beam.

A seaworthy ship must have a strong hull and yet, to be economic, it must not be weighed down with excessive structure. The hull must be so shaped and proportioned as to make the ship "sea-kindly." By this we mean that it can continue operations without excessive motions or water on deck despite adverse weather conditions.

A ship must be self-sustaining to some degree. It must be able to carry enough fuel, food, and fresh water to enable it to operate as an independent unit for some period of time. That may be a matter of minutes for a cross-river ferry, or a matter of months for a naval submarine. To be self-sustaining, a ship and all its components must incorporate a high degree of reliability. There are no handy repair shops in mid-ocean.

A ship must be controllable as to speed and direction. It must have at least reasonably comfortable accommodations for the crew with proper attention paid to food service, off-duty amenities, and freedom from disturbing noise, vibrations, motions, or extremes of temperature. It should have user-friendly control and monitoring equipment in carefully laid-out working spaces.

A ship must be safe and especially so if passengers are to be aboard. Ways must be found to minimize the chance of damage, whether from fire, explosion, storm, collision, or grounding. Recognizing that such damages may occur, our ship should incorporate features that will act to save it from complete destruction. Failing that, various forms of lifesaving gear must be provided so as to save lives even if the ship itself is lost.

Success in moderating all of the foregoing conflicts can follow only if the naval architect always keeps in mind the vessel's central function. Take for example a cruising yacht. Such a boat must above all provide a comfortable accommodation for the owner and guests. (This implies enough area for not-too-crowded private cabins, social centers, and food service functions.) It must have an appropriate degree of self-sufficiency. None of those other considerations can be overlooked, but compromises must always be slanted toward comfort and self-sufficiency.

3. *Basic Definitions*

Before going further I must take time to explain some technical terms that you will want to understand when talking to a naval architect. You will find brief definitions of many technical terms in the

Glossary. Also, in case you missed it, there is a list of abbreviations just before the table of contents. But let us look at a few basic expressions right now. (Be warned, however, that these are in many cases greatly simplified. They would certainly not hold up in court.)

First of all, what is the difference between a ship and a boat? Generally speaking, a ship is big, a boat is small. There is no exact dividing line, and inconsistencies abound. For example, on the Great Lakes thousand-footers are called "ore boats." Other pertinent terms are *vessel* (any size) and *craft* (usually small). Please note that the plural of *craft* is also *craft*. (The term *crafts* applies to handmade articles.)

We call the front end the *bow*, the back end the *stern*. *Forward* means toward the bow; *aft* means toward the stern. We call the leading edge the *stem*, never the *prow*.

Port and *starboard* mean the vessel's left and right sides, respectively. Using those terms instead of left and right makes sense. If you are standing in the bow looking aft and refer to "the left lifeboat," who knows what you mean—your left, or the ship's? Want a hint on how to remember which is which? *Port* and *left*: each word has only four letters.

Draft is the vertical distance from the water level down to the lowest point of the ship's hull. *Depth* is the vertical distance between the uppermost continuous deck and the bottom of the hull at midlength. *Freeboard* is approximately the difference between the two.

So now I had better define *deck*. It is a part of the ship's structure corresponding to a floor (or flat roof) of a building. *Bulkheads* correspond to walls. The *shell* comprises the watertight plating around the sides and bottom of the ship.

The *innerbottom* is a watertight interior skin, usually separated from the bottom shell by a shallow crawl space. The structure within that space plus the innerbottom plating itself form what is called the *doublebottom*.

We must not overlook several weight definitions. In the United States, the units of weight are almost always long tons (2240 lb). The exception is found among inland waterways where the short ton (2000 lb) is more commonly used. In most other countries, the unit of weight is the tonne (1000 kg). This is usually pronounced "tun" although some naval architects pronounce it to rhyme with "funny" so as to differentiate it from the long ton. A tonne weighs about 2205 lb. Some people wonder why an illogical sounding figure like 2240 lb came to define the long ton. Simple enough: It dates back to 15th century England. A major import then was wine and it arrived in oversize barrels, each of which (when full of wine) weighed 2240 lb. Simple as that.

Light ship is the vessel's empty weight with minor increments for spare parts, etc. *Deadweight* is the ship's lifting capacity, made up of cargo, fuel, fresh water, crew, passengers and their belongings, food, and other expendable supplies. The *displacement* is the total weight and equals the sum of light ship and deadweight.

In naval parlance, the *payload* is the difference between the full load displacement and the weight of the empty ship devoid of any military equipment such as weapons, detection gear, communications gear, etc.

Tonnage (more correctly *registered tonnage*) applies to legally defined approximations to a ship's size (gross tonnage) and cargo hold capacity (net tonnage). The units have nothing to do with weight but are given in terms of 100 cubic feet.

A *nautical mile* is taken as one minute of arc along the equator and is equal to about 6080 ft. That makes it close to 15 percent longer than a statute mile (5280 ft).

Speed is measured in *knots*, which are nautical miles per hour.

The term *scantlings* does not refer to any specific part of a ship. It is a generic term pertaining to the materials from which the ship is built. Scantlings generally define the cross-sectional characteristics of structural members but say nothing about the length of the item or whether it is curved or straight.

In way of is frequently heard. It means "in line with" or perhaps "next to," or "at the intersection of."

Classification societies are (generally) independent organizations to whom merchant shipowners can turn for guidance in designing, building, and maintaining seaworthy hulls, machinery, and equipment. Without a certificate of classification from such a society, an owner would have trouble acquiring insurance on the ship.

Ratings are seafarers who are not officers.

To complete this brief lexicon, here are some miscellaneous matters of semantics and maritime culture.

Are boats and ships *it* or *she*? By tradition, specific ships are usually referred to as *she*, or, in the possessive, *her*. In technical writing or in speaking of ships collectively, we are just as likely to treat them as neuter objects: *it* or *its*.

To a layman a vessel's *wake* is the collection of waves it leaves astern. To a naval architect, *wake* refers to the body of water that is dragged by friction and tends to follow the ship as it moves through the water.

Seacocks are large valves that can be opened to the sea so as to make it easy to scuttle (that is, sink) the ship. Aside from some military craft, they exist only in the minds of fiction writers.

The term *tall ship* (derived from the famous Mansfield poem) is a journalist's overused cliché. The former cliché was *windjammer*. Most of us prefer to call ships of that ilk what they are: *training ships* or perhaps *square-riggers*.

We never go *downstairs* on a ship, only *below*.

What most laymen call a *dock* is more correctly called a *pier* or *finger pier*. A *dock* or *slip* is the body of water between two piers. A *quay* (pronounced "key") is a masonry structure usually built along the shore. A *wharf* may be either a pier or a quay.

Further Reading

Lewis, Edward V. and O'Brien, Robert, *Ships*, Life Science Library, Time, Inc., New York, 1965.

Haws, Duncan, and Hurst, Alex A., *The Maritime History of the World*, Teredo Books, Ltd., Brighton, U.K., 1985.

Haws, Duncan, *Ships and the Sea: A Chronological Review*, Thomas Y. Crowell and Company, Inc., New York, 1975.

Casson, Lionel, *Illustrated History of Ships and Boats*, Doubleday and Company, Inc., Garden City, N.Y., 1964.

Fig. 1.1 Fireboat. This 78-ft boat was built at Grafton Boat, Grafton, Illinois, for the Massachusetts Port Authority. It has a pumping capacity of 6800 gallons per minute, and is designed for both harbor and airport service. Courtesy John W. Gilbert Associates, Inc.

Fig. 1.2 Vessels in a range of sizes. Upper left: Cruise ship *Sovereign of the Seas*, offering weekly sailings out of Miami. Courtesy Royal Caribbean Cruise Line. Upper right: W.E. Zimmie's yacht *Endless Summer*. Courtesy Eileen Zimmie. Lower left: Fishing trawler *Victor*. Courtesy John W. Gilbert Associates, Inc. Photo by Herb Reynolds. Lower right: Launch of stern trawler *Ranger*, a 120-ft commercial fishing boat operating on the Grand Banks off Newfoundland. Courtesy John W. Gilbert Associates, Inc.

Fig. 1.3 Naval vessels. Upper: Aircraft carrier CVN 71; *Teddy Roosevelt* on sea trials in October 1986. Courtesy Curtis Woolard, Jr., Newport News Shipbuilding and Dry Dock Co. Photo by Stu Gilman. Lower: Submarine SSN 750 on sea trials. Courtesy Curtis Woolard, Newport News Shipbuilding and Dry Dock Company. Photo by Caroline Kiehner.

Fig. 1.4 Two powerful vessels. Upper: A high-speed containership. Container boxes are carried both above and below decks. Courtesy C. R. Cushing and Co. Inc. Lower: A quadruple-screw river towboat. These shallow-draft boats are designed to push barges lashed together into what are called flotillas. Photo by *St. Louis Post Dispatch.*

CHAPTER II

GENERAL ARRANGEMENTS

1. *Basic Principles*

No matter what kind of ship or boat is under consideration, there are certain common sense rules that a naval architect will want to remember when deciding where to locate each of the various component parts. The first thing to realize is that there will inevitably be conflicts. (For example, in a small power yacht the engine almost invariably seems to foul up the arrangements by occupying a space right near the middle of the cabin accommodations.) Priorities must be established and compromises reached. The naval architect's rankings will be in one order on a cargo ship and another on a yacht. Moreover, there will always be a need for subjective judgment and no two designers will come up with the same answers. Like so much in naval architecture, arrangement decisions involve art as well as science, and it's the art that makes the subject so fascinating.

Computer algorithms and graphics may be of great help in roughing out arrangements, but the actual decisions are best left to the naval architect.

The following principles must all be kept in mind throughout the task of deciding what is to go where in any ship or boat. By this I mean that naval architects try to satisfy *all* the requirements simultaneously, not just one at a time. Even with this kind of guidance, there is in the final analysis no substitute for experience and a natural flair for creative work.

1. The ship's primary function must never be overlooked when settling conflicts.

2. Those design features that ensure success in meeting the primary function must govern the design. For example, in a commercial fishing boat, the designer should start by selecting and locating the fish-catching equipment. In a cargo ship, the first task is to find out the form in which the cargo is to be moved. Then decide how the cargo is going to be moved on and off the ship, through what kind of openings in the hull, and where it is to be carried within the hull. Having made those key decisions, the designer will wrap the design around them. You can judge for yourself what the key decisions should be in other kinds of floating craft.

3. Nothing is to be gained by resisting the inevitable. The forepeak bulkhead and forepeak just *have* to go up there at the forward end. Decisions like that are easy. (If you need help with the terminology, see the Glossary.)

4. There is benefit in finding locations where the hull form fits the function. Liquids can be carried in oddly shaped spaces; cargo container boxes cannot.

5. The big items should be located first. When you are ready to load your car to start on vacation, common sense tells you to load the big packages first and then tuck the smaller items in around them. The same applies to arranging spaces on ships or boats.

6. The comfort and well-being of the passengers and crew merit high priority. Keeping passengers happy is too obvious to require further discussion. But the welfare of the crew must also rank high if the naval architect wants the ship to be operated effectively and good things to be said about the design.

7. Stability and trim must always be kept in mind. What happens during the voyage as fuel is used up? When no cargo is available, is there enough ballast capacity to put to sea safely?

8. Things that ought to go together should be consolidated wherever possible. The galley (kitchen) should be placed next to the eating spaces and also close to the food stores. To minimize the extent of piping, wiring and vent ducts, accommodations should be placed close to the machinery space. (In a large ship the accommodations are usually placed right over the machinery space.) Sanitary spaces for adjacent cabins are best situated back-to-back to minimize the amount of piping.

9. Things that clash should be segregated. A noisy mooring winch does not belong atop somebody's cabin. The library should be kept well away from the disco bar.

10. Major structural components should be integrated into the general layout. Placing a pillar in the middle of a passageway is clearly bad form.

11. In the engine room, the clearances required for effecting repairs must be kept in mind. The naval architect must try to provide enough vertical clearance over any large diesel engine so that repair crews can pull the pistons. Similarly, clear space should be reserved to allow pulling tubes out of heat exchangers. And, room must be allowed so a human being can get at all the equipment with wrench and screw driver.

12. The layout should, in general, be conceived from the inside out. By this I mean that the naval architect will usually settle on the interior arrangements (shown in what we call an *inboard profile*) before worrying too much about exterior appearances. In a yacht or passenger ship, one may of course want to reverse the procedure.

13. The designer should always try to imagine the ship or component in service. How will the crew manage to make things go? What about conditions in bad weather with seas running, wind blowing, rain falling, boat rolling and pitching? Would the designer like to be aboard?

2. *Sample Illustrations*

Figures 2.1 through 2.5 show arrangement profiles of five widely differing kinds of boats or ships. Figure 2.6 shows silhouettes of the five vessels all drawn to the same scale so as to give you an idea of relative sizes. In these profiles you may note that all bows are drawn to the viewer's right, a strongly held maritime drafting convention. Figure 2.7 shows the exterior arrangements of an oceanographic research ship and gives you an idea of the extensive deck gear appropriate to such operations.

Further Reading

Kiss, Ronald K., "Mission Analysis and Basic Design," Chapter I in *Ship Design and Construction*, Robert Taggart, Ed., Society of Naval Architects and Marine Engineers, Jersey City, N.J., 1980.

Michel, Walter H., "Mission Impact on Vessel Design," Chapter II in *Ship Design and Construction*, Robert Taggart, Ed., Society of Naval Architects and Marine Engineers, Jersey City, N.J., 1980.

Tapscott, Robert J., "General Arrangement," Chapter III in *Ship Design and Construction*, Robert Taggart, Ed., Society of Naval Architects and Marine Engineers, Jersey City, N.J., 1980.

Baxter, Brian, *Naval Architecture: Examples and Theory*, Charles Griffin & Company, Ltd., London, 1967.

Munro-Smith, R., *Elements of Ship Design*, Marine Media Management, Ltd., London, 1975.

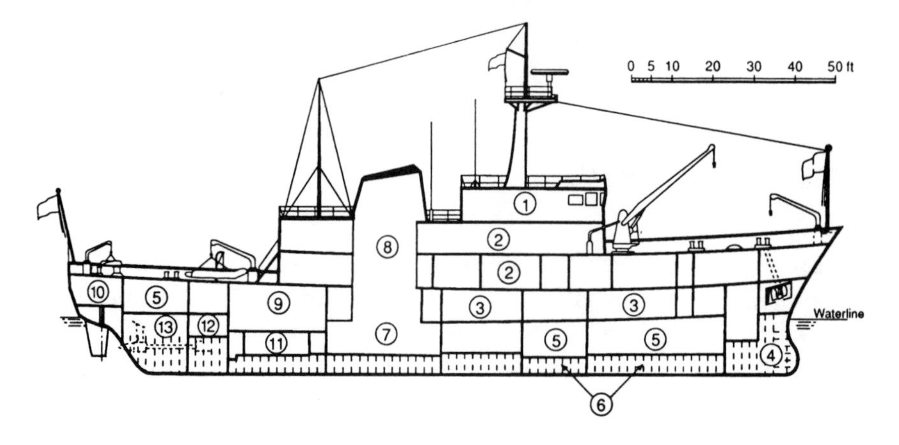

1 Navigating bridge
2 Public spaces, food service, scientific equipment, etc.
3 Accommodations
4 Forepeak, used for water ballast
5 Storage spaces
6 Doublebottom, used for carrying fuel oil
7 Engine room
8 Engine casing, provides room for exhaust system, access, and hotel service systems between engine room and accommodations.
9 Auxiliary engine room: electrical generators, etc.
10 Steering gear room
11 Fuel oil tank
12 Potable (drinking) water tank
13 Afterpeak, used for water ballast

Fig. 2.1 Oceanographic research ship. This little ship is used for coastal survey work. The overall length is just under 190 ft (58 m) and the speed is 12 knots.

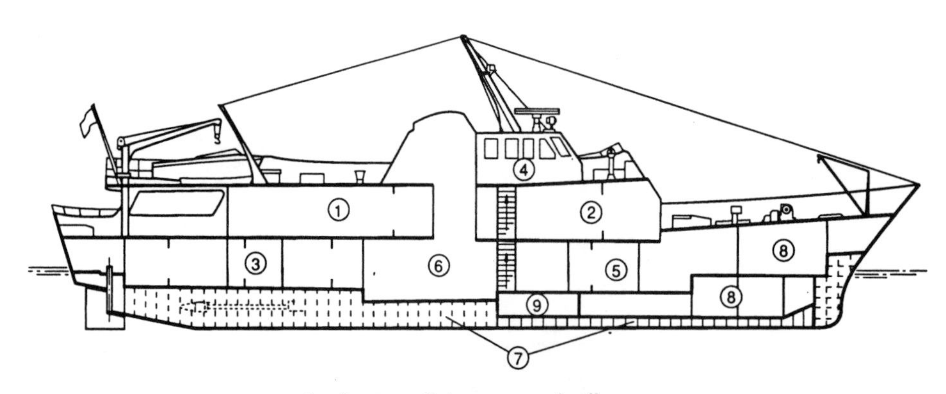

1 Lounge, dining room and galley
2 Owner's stateroom
3 Guest accommodations
4 Navigating bridge
5 Crew accommodations
6 Engine room
7 Doublebottom, used for fuel storage
8 Storage space
9 Potable water tanks

Fig. 2.2 Motor yacht. This is a true luxury yacht, intended for extended ocean cruising at 13.5-knot speed. The vessel can continue for 3600 miles before replenishing fuel or supplies.

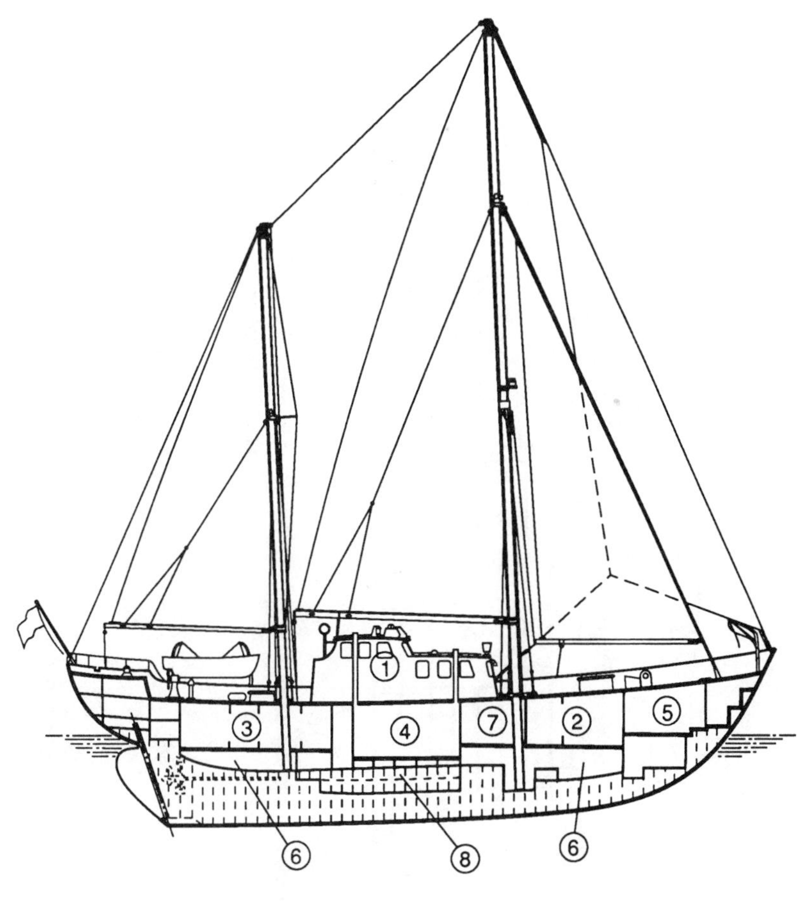

1 Lounge, dining room and navigating space
2 Owner's stateroom
3 Guest accommodations
4 Engine space
5 Crew accommodations
6 Storage spaces
7 Galley and food stores
8 Fuel tanks

Fig. 2.3 Sailing yacht. This 110-foot (34 m) ketch-rigged boat has twin-screw diesel engines for auxiliary power.

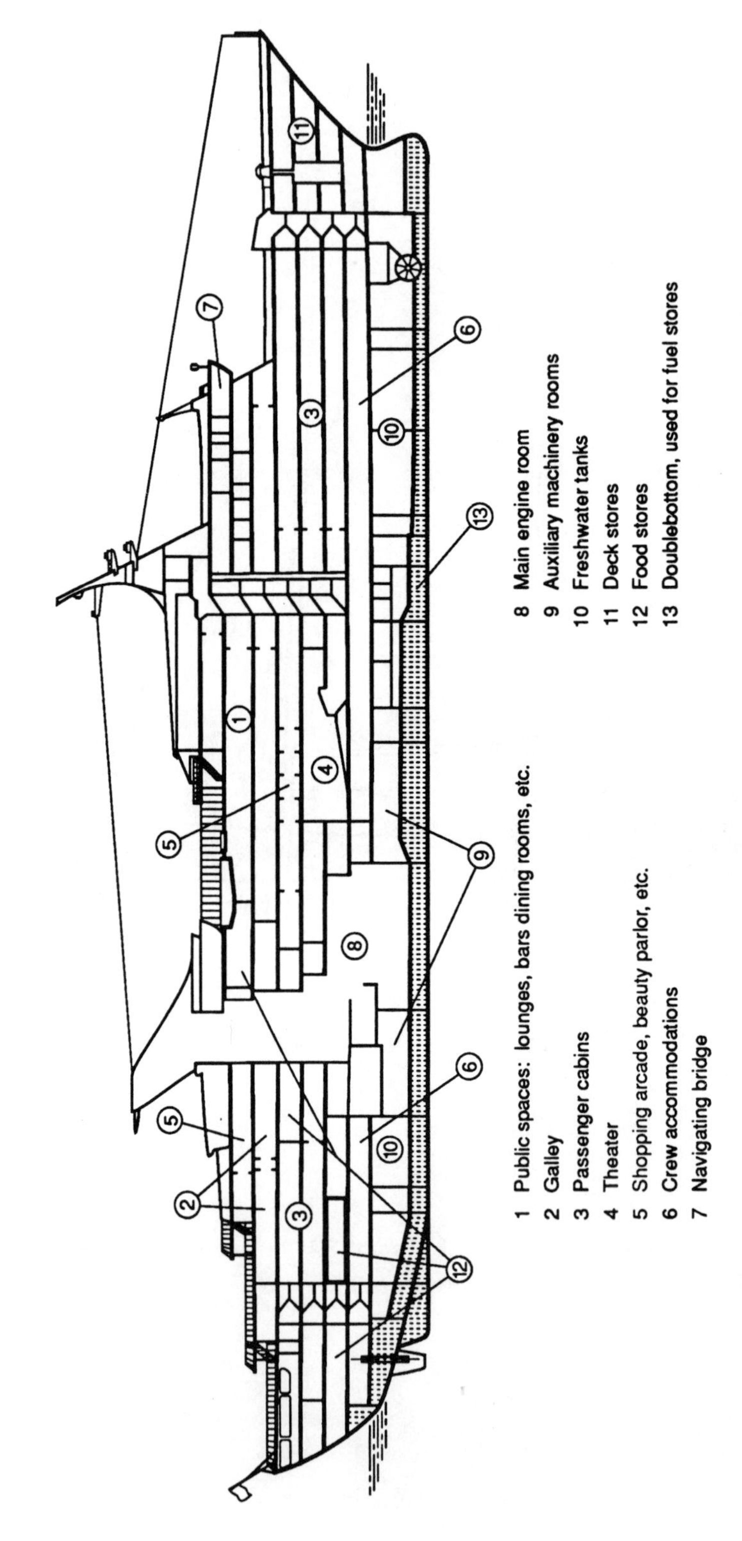

Fig. 2.4 Cruise ship. With an overall length of 525 ft (160 m), this floating resort hotel carries 700 passengers on two-week cruises.

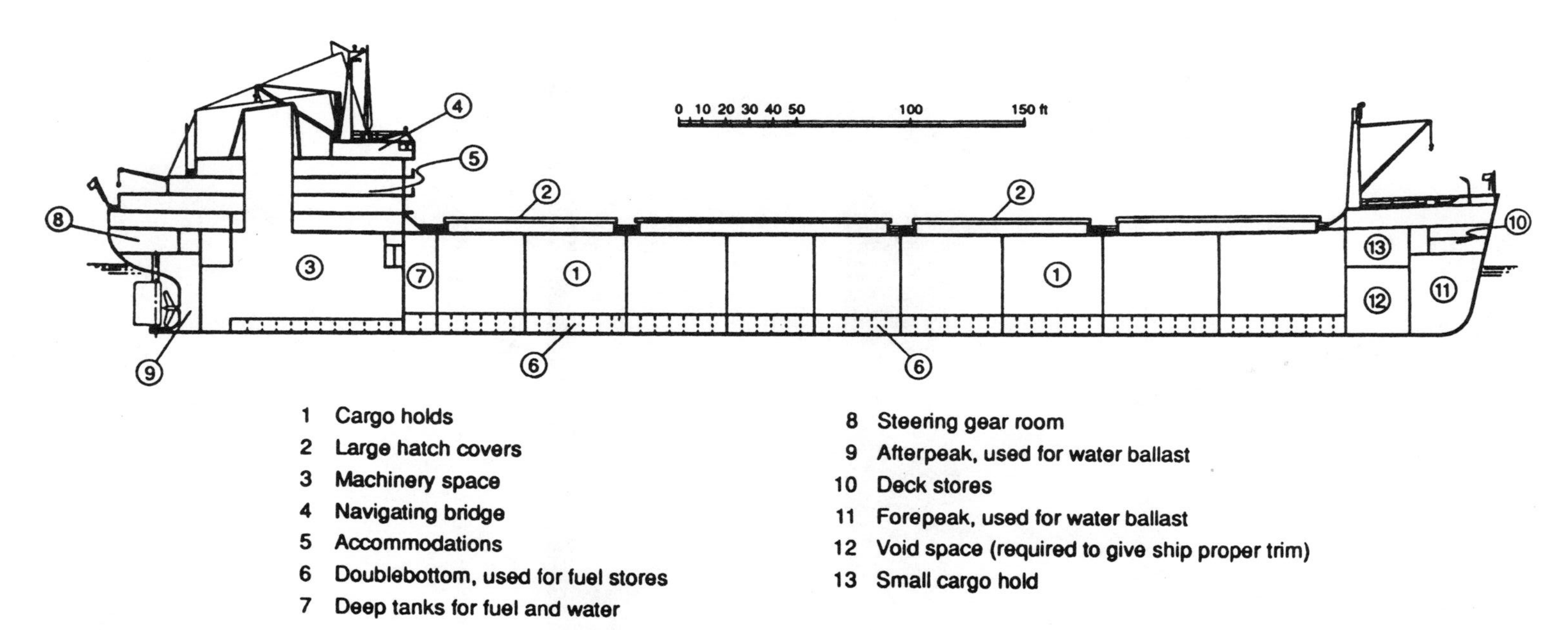

Fig. 2.5 Dry-bulk cargo carrier. This ship carries 27 000 long tons of grain or iron ore at a speed of about 14 knots.

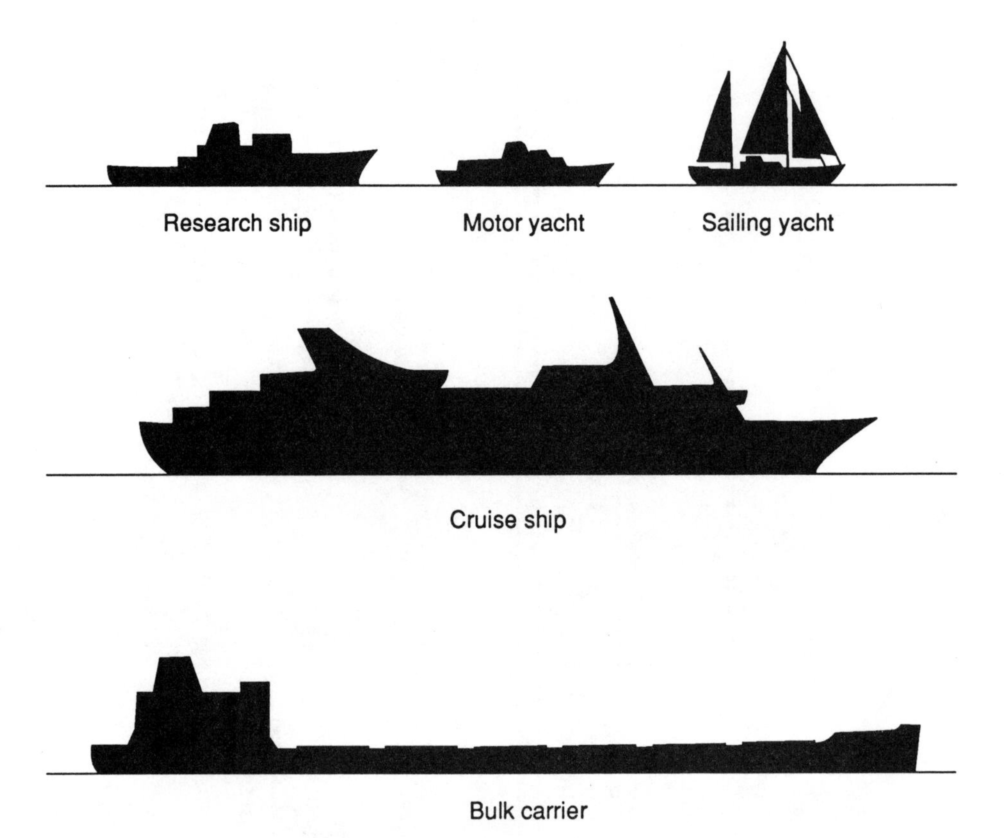

Fig. 2.6 Comparative sizes. These sketches will give you some idea about how our five typical vessels compare in size.

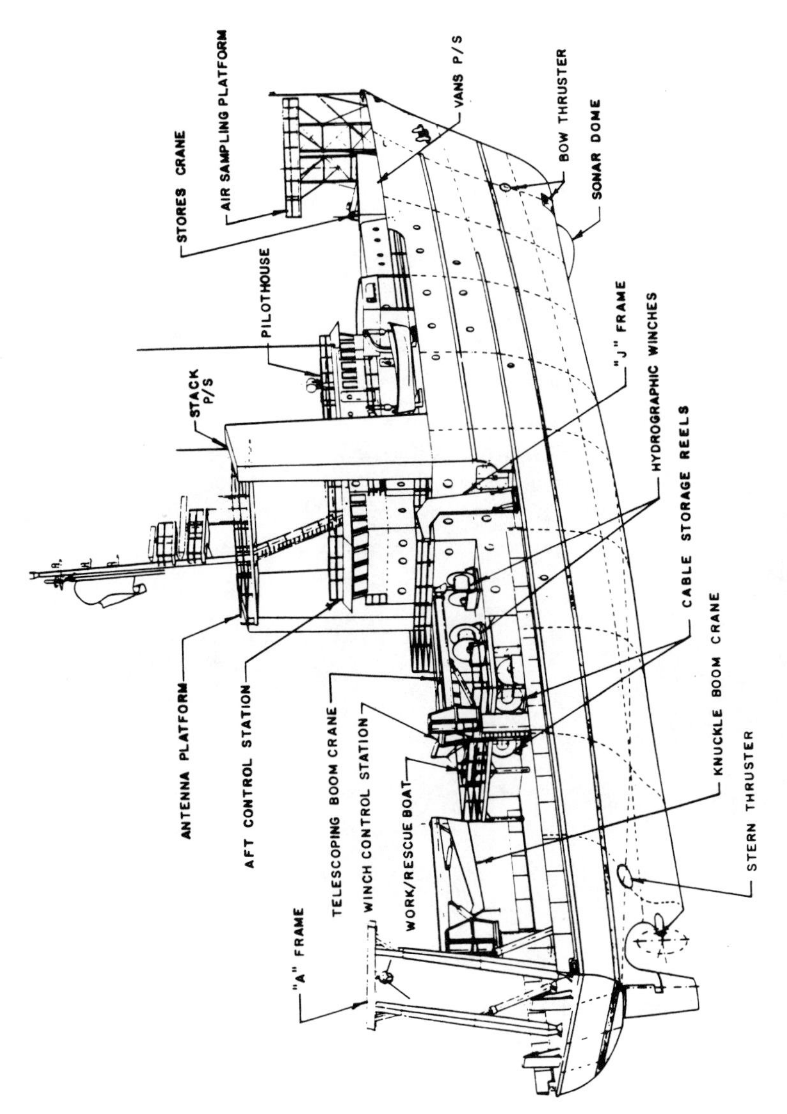

Fig. 2.7 Deck arrangements on an oceanographic ship. Courtesy M. Rosenblatt and Son.

Fig. 2.8 Coastal passenger ship *Provincetown.* This 185-ft ship was built in 1979 at Jakobson Shipyard in Oyster Bay, New York. Courtesy John W. Gilbert Associates, Inc.

Fig. 2.9 Two kinds of bulk carriers. Upper: Tanker *Esso Le Havre*, of 381 000 deadweight tons. This was the first vessel of such an extreme size to transit the Suez Canal (in 1981). Courtesy R.F. Klausner, Exxon Company, International. Lower: Self-unloading collier *St. Clair*, typical of Great Lakes bulk carriers. Photo by Harmann Studios, Sturgeon Bay, Wisconsin.

Fig. 2.10 Stern scalloper *Huntress.* This 104-ft commercial fisherman was built by Edward Gamage Shipbuilding of East Boothbay, Maine. Courtesy John W. Gilbert Associates, Inc.

CHAPTER III

WORKING AREAS AND ACCOMMODATIONS

1. Human Factors in Design

Throughout the design of a ship or boat of any size, a naval architect should never forget the human beings who will be aboard. One must keep in mind the many factors that will influence the crew's ability to work at top efficiency and to be willing to sign on again for future voyages. Clearly, if the design is for a yacht or passenger ship, these human factors demand even stronger consideration. The importance of these matters is so great that the prudent naval architect will always try to bring the affected parties into the decision-making process at nearly every stage of the design.

1.1 Ground rules. Depending on the kind of ship and owner, there may be a body of basic rules that must be met. In a U.S.-flag cargo ship, for example, there are Coast Guard rules about minimum crew size and accommodation standards. There will also probably be labor-union agreements. The designer must understand all these basic constraints without, however, necessarily treating minimum requirements as maximum design standards.

Early in the design stage the designer should learn from the owner as much as possible about the working conditions (for example, two-watch versus three-watch systems) and social conventions to be expected on the proposed vessel. How much off-duty mixing between officers and ratings is appropriate? (They may want to be segregated at meals but might well be willing to share a common swimming pool.) Certainly, also, the designer must know from the start how many seafarers of various ranks are to be aboard and what their work assignments are to be.

1.2 Social/psychological considerations. Naval architects must remember that shipboard life is highly unnatural, so a wide variety of human-relations problems is likely to arise. Although a naval architect cannot solve all such problems, care in design can help. For example, as we move toward ever-smaller crews, loneliness becomes more of a problem. The designer should therefore try to arrange the accommodations in such manner as to increase the likelihood of people making face-to-face contact as they move about.

1.3 Automation. In parts of the world where high wages prevail, there are economic incentives to increase crew productivity so as to allow safe and efficient operation with reduced crew numbers. Developments in this direction are usually associated with shipboard automation. There are, however, many other steps that can be taken to achieve such economies. More maintenance work can be done ashore or by roving repair crews. Components can be selected with a greater emphasis on durability and reliability. Corrosion-resistant materials may be specified. Perhaps most important of all is the move toward better organization of work, with versatile crews who are willing and able to help as the need arises in either engine or deck departments.

All such matters of manning practice should be clearly understood before the design of working areas and accommodations begins.

2. *Working Areas*

2.1 Aim. In designing the working areas on a ship the basic aim is to provide a physical working environment that will allow the crew to perform the work effectively and ensure a reasonable degree of safety and comfort. There must be adequate control of noise, temperature, vibrations, ventilation, odors, and illumination. Some elementary understanding of ergonomics (man-machine systems) is also essential in the design process. The following subsections describe three working areas to which the naval architect devotes particular attention: the navigating bridge, the food-service arrangements, and the machinery spaces.

2.2 Navigation and control. In a ship of any size, the navigation and control functions are accorded a special work area called the *wheelhouse* or *bridge*. Figure 3.1 shows a typical layout for such a control center. The bridge incorporates such desirable features as 360-degree visibility, windows sloped (outward at top) to eliminate reflections from interior lights, and enclosed areas for shelter, but also open areas for maximum visibility and hearing. It includes such conveniences as a toilet, and a sofa on which the captain can rest during prolonged periods on the bridge. The chart table is positioned so that the person using it faces forward; this saves having to shift mental gears when visualizing the ship's actual course. As much as possible the various bits of equipment are located so that the user is facing forward. The bridge wings extend a few feet beyond the ship's side. This allows the master or mate a clear view of quay, lock wall, or tugs alongside. While it is not apparent on the drawing, the vertical location is high enough to allow good visibility over the bow.

2.3 Food service. Tasty, nutritious food served in pleasant surroundings is an important part of the implied social contract between shipowner and crew in any fleet that aspires to long-term retainment

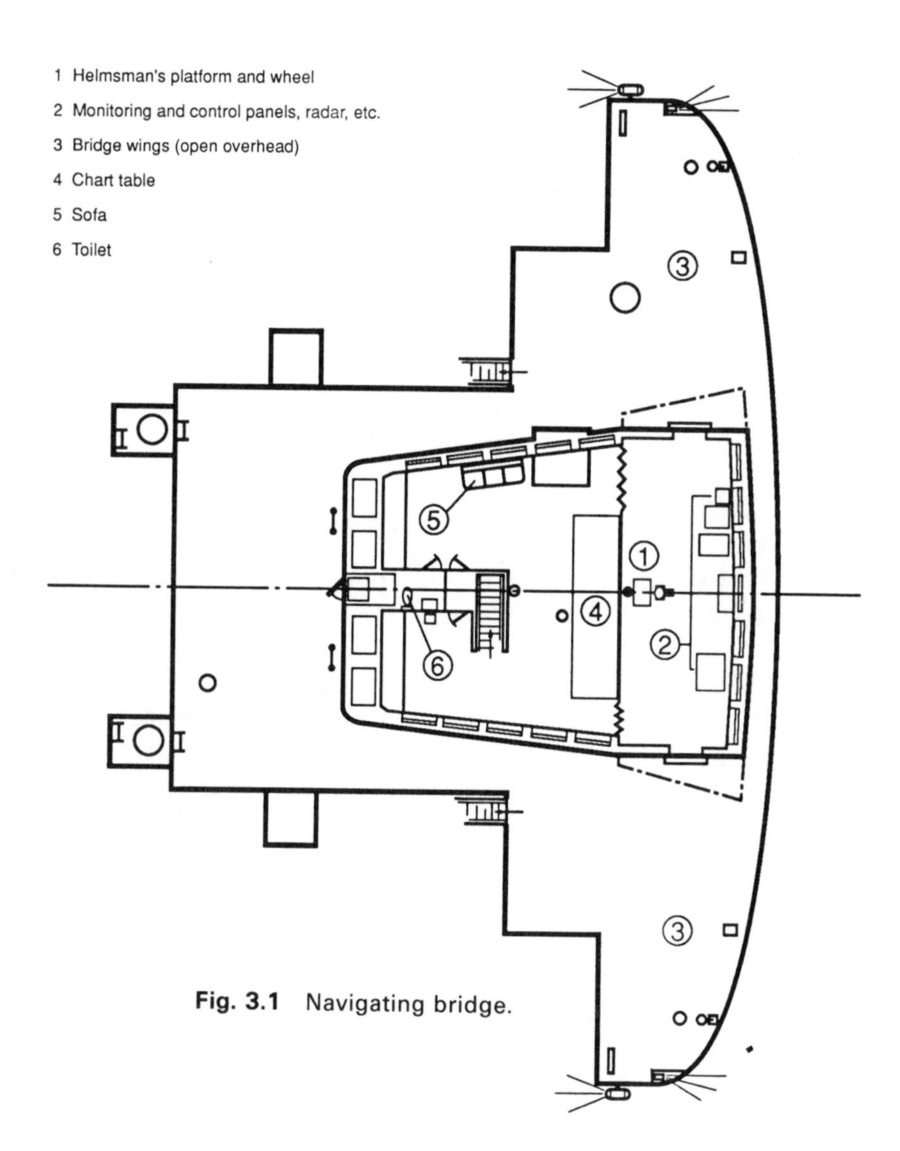

Fig. 3.1 Navigating bridge.

of personnel. This is particularly true where restrictions on accommodation area preclude luxurious amenities. As one crew member put it, "If the food is good, we can handle the rest." Two basic questions that must be answered right away are how many separate messing spaces must be provided and what are their seating capacities? In a small commercial vessel such as a tugboat, space constraints usually dictate a single mess for both officers and ratings. On large ships there may be a considerable degree of segregation, with individual eating areas for deck and engine crews and separation between officers, petty officers, and ratings. Thus there may be up to six different messes plus a special area set aside for duty officers in dirty clothes.

There are few aspects of shipboard life that carry such a high degree of brass-bound tradition as that associated with food service. Naval architects try their best to learn exactly what is expected here.

Having settled on the number and sizes of messing spaces, the designer tries to position them adjacent to the culinary working areas (galley, scullery, and garbage disposal). Storage compartments for dry and refrigerated food should also be positioned near the galley. Care should be given to ease of operation: loading stores aboard ship, moving them to the galley, serving the crew, and disposing of the garbage. Similarly, the galley equipment should be selected and located with the aim of allowing efficient operation.

There is also the matter of safety in the galley. Ranges, for example, should be oriented athwartships (see Glossary) so that a sudden roll of the ship will not deposit a kettle of boiling soup on some unsuspecting cook.

Figure 3.2 shows a typical arrangement of the food service spaces on a merchant ship.

2.4 Machinery spaces. There are two parts to the task of settling on machinery arrangements. The first is to decide where to locate the engine room. The second is to fix the details as to what goes where. In the interest of brevity we shall consider only the first part.

In deciding where to place the machinery in a cargo ship, we must defer to the primacy of the ship's central function: that of transporting cargo. There are many other elements, but that is the first. In containerships, dry-bulk carriers, and tankers, the nature of the cargo more or less dictates that the engine room should be placed at or near the

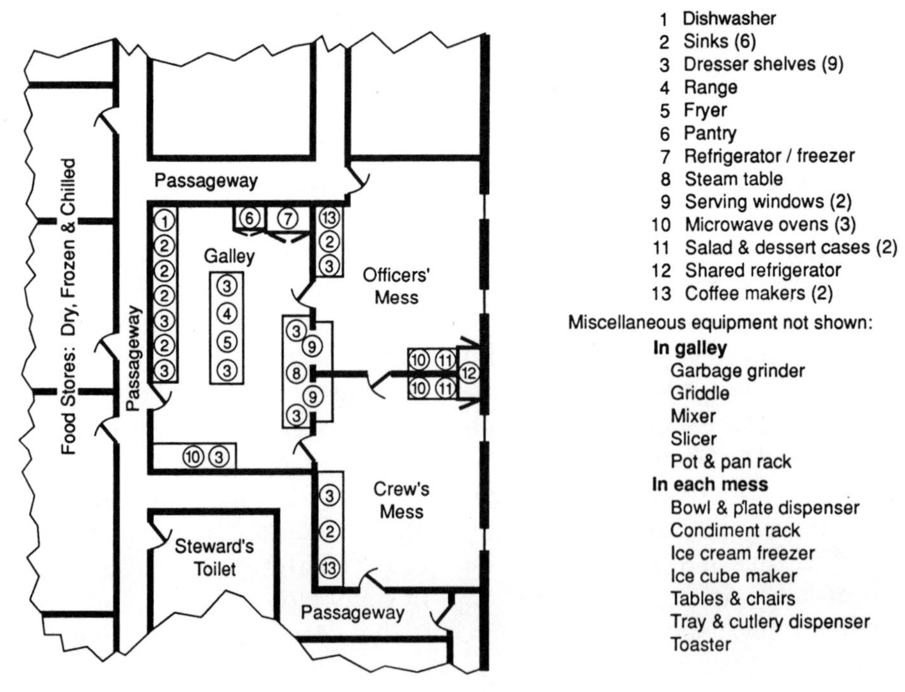

Fig. 3.2 Arrangement of food service spaces on a merchant ship.

stern. In roll-on/roll-off (RO/RO) ships, ferries, or break-bulk cargo ships, there are good reasons to place the machinery near amidships.

The midship location has several advantages. Because of a more even longitudinal distribution of weights, the ship can be built with a somewhat lighter structure than would be the case with machinery aft. (You will learn why in Chapter XII.) It is also easier to eliminate the problem of undesirable trim that complicates the design when machinery is in the stern. You will notice that tankers, for example, usually incorporate an empty hold near the bow to balance the relatively light weight of the engine room in the stern. Without that empty hold, the tanker would have an unwanted trim by the bow. As mentioned in Chapter II, the accommodations are usually placed vertically in line with the machinery space. This means that with machinery amidships there would be far less discomfort from pitching and vibrations in the quarters than would be the case with the machinery aft.

The midship location, however, usually brings with it the problem of running a long propeller shaft to the stern. That shaft is enclosed in a steel tunnel that occupies valuable space and interferes with cargo stowage.

A key consideration in all this is the matter of total engine room size. This decision contains the seeds of friction between those who design the machinery spaces and those in charge of overall design. In a minority of cases, such as ferries, there may be plenty of room. In most ships, however, economic pressures place constraints on the extent of the machinery spaces. In principle, then, naval architects want to make the engine room or rooms as small as is reasonably possible. This means that the machinery components must be placed just far enough apart so that the crew will have elbowroom to service, inspect, and on occasion repair each item of hardware.

In the case of direct-connected diesel engines located in the stern, the governing size constraints may be the length and height of the main engine. There may be space to waste port and starboard.

The ship designer must also think about arrangements for supplying air to the engine room and getting rid of exhaust gases. Then there is the matter of distributing the various hotel services (such as ventilation ducts, electrical wiring, and piping systems) between the engine room and the accommodations. In addition, the designer must remember to provide ladders and gratings to allow access to, and within, the engine room. To make space for all these functions, a large vertical air shaft running from the top of the engine room right up through the accommodations and other deckhouse spaces is usually provided. This is called the *engine casing*. Because the number and sizes of the various systems progressively diminish in the higher levels, the horizontal cross-sectional size of the engine casing can be correspondingly reduced. And so we usually find the casing configured in stepped pyramid style. At the uppermost deck the engine exhaust enters the bottom of the smokestack and there may be a skylight or other removable closure than can be opened to allow easy removal of large items of machinery.

The foregoing discussion has been directed at cargo ships. Similar common sense thinking is used in all the other kinds of vessels, large or small.

Chapter X, Choosing Propulsion Machinery, has additional things to say about machinery arrangements.

3. Accommodations

Human beings need alternating periods of privacy and social contact. Privacy is afforded in crew cabins; social contact comes about during work periods but also, more importantly, in various public spaces during off-hours. These spaces include mess rooms (discussed above) and perhaps other social gathering places such as lounges and game rooms. There may also be spaces, such as swimming pools and gymnasiums, set aside for off-duty physical activities.

One of the first steps in designing the accommodations is to establish the fore and aft extent of the deckhouse. In containerships, where deck area is at a premium, the accommodations may have to be stacked up in a high-rise configuration. That adds to the problem of providing convenient access, but often cannot be avoided.

3.1 Private quarters. In general, all crew cabins should be located above the deepest likely operating waterline, and all should be "outside" rooms. By this I mean they should be fitted with windows or ports allowing an outside view. Neither of these constraints applies to passenger cabins, some of which may be located far below the waterline and which would have no port or window. Needless to say, such undesirable features apply only to the less expensive accommodations.

In the old days the ratings' sleeping quarters were crammed into the fo'c's'le (forecastle), but this is no longer allowed. The U.S. Coast Guard, in looking after the seafarer's safety, has decreed that no one be berthed in the extreme bow (a portion of the ship much inclined to be intimately involved in 50 percent of all collisions).

Crew's quarters should, as much as possible, be located close to the place of work. They should also be close to the public rooms. Thus we see that engine ratings are usually given cabins on a fairly low deck. Above that level would be a deck for the food service functions. The deck crew ratings might be on the next deck above, the officers of all departments above that, and the navigation spaces highest of all.

Crew cabins need be only large enough to attend to the user's private needs: sleeping, reading, and sanitary functions, plus reason-

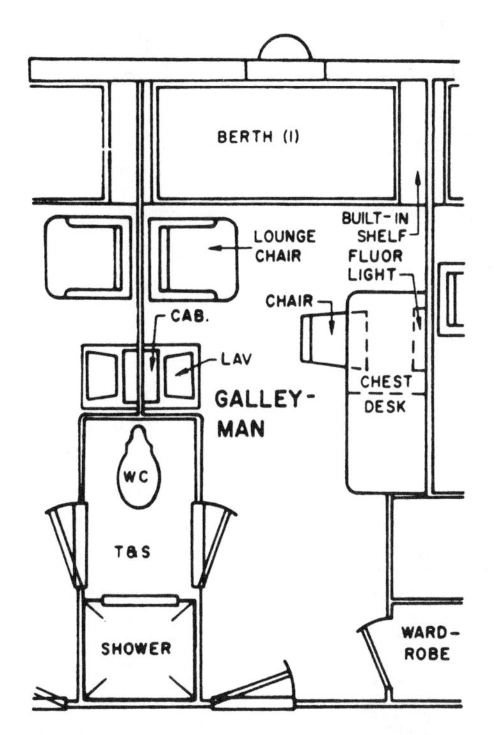

Fig. 3.3 Typical layout for a rating's cabin on a merchant ship (source: Tapscott, 1980).

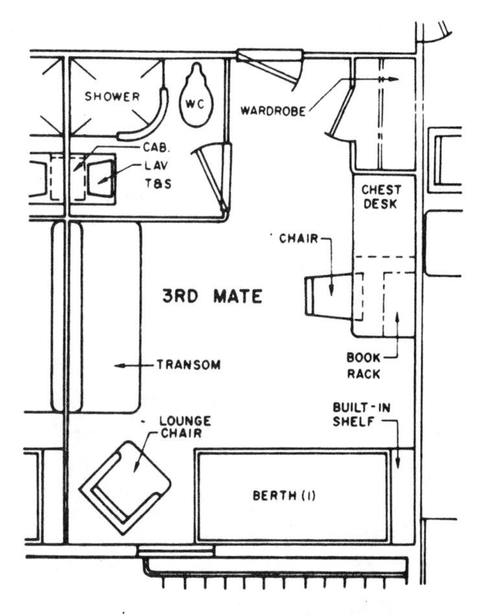

Fig. 3.4 Typical layout for an officer's cabin on a merchant ship (source: Tapscott, 1980).

able clearances for housekeeping. Figure 3.3 shows the layout of a typical single-occupancy cabin for a rating on a merchant ship. Note that the toilet/shower is shared with the adjacent cabin, but each has its own washbasin (lavatory). Figure 3.4 shows the layout for an officer's cabin. Here you see a private toilet/shower (T&S) space and a transom (settee) as well as a berth. On saltwater ships, there is a strong tradition that berths must be oriented fore and aft, whereas some seamen on the Great Lakes subscribe to an old tradition that berths must be oriented athwartships. Settees, where fitted, should be at right angles to the berth; then, no matter how the ship may roll or pitch, the occupant may hope to find a comfortable place to sleep.

You have been reading so far about relatively large ships. On small craft, accommodation standards must of necessity be compromised to some degree. In a cruising boat, for example, you will usually find all interior accommodation functions jammed into one or two highly cramped spaces (see Fig. 3.5). A good deal of clever design goes into making such spaces both comfortable and practical. This work is doubly challenging because the hull into which the features must be crammed presents no flat, rectangular surfaces.

3.2 Public spaces. Dining facilities have already been considered in our section on working spaces. In some small merchant ships, those spaces form perhaps the only public gathering places. On larger ships, however, there may be any number of other public off-duty amenities: swimming pools, video TV rooms, gymnasiums, and hobby shops. Such extra features are particularly appropriate on ships engaged in long ocean voyages. Boredom is often a serious problem on such ships, and it is likely to be intensified where short port times preclude a change of pace for the seafarer.

Since the public spaces in a merchant ship provide perhaps the sole setting for socializing, they should be made physically inviting to enter and should be arranged to encourage the formation of conversational groups.

In passenger ships, of course, a great deal of emphasis goes into the design and furnishing of all public spaces. Professional interior designers are often brought into the process because in many ways these public spaces form the very heart of the arrangement and have the strongest effect on the passengers' perception of the ship.

4. Access

4.1 General. Good access about a ship requires careful planning from the earliest stages of laying out the accommodation and working spaces. Proper access is a factor in both segregation and integration. One of the first steps in blocking out living and working spaces should

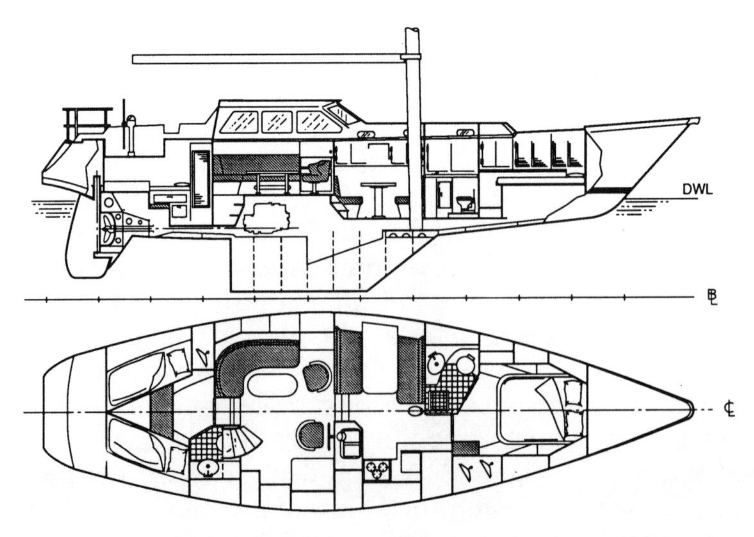

Fig. 3.5 Inboard profile and arrangement plan for a 47-ft cruising yacht. Courtesy of Jay Benford.

be to locate corridors, stairs, doors, and elevators. Naval architects have to give thought to traffic flow between cabins, work places, and public spaces. Except in the smallest vessels, an attempt should be made to provide at least two means of getting to the open deck and life-saving equipment from any place on board. Wherever practical, sheltered access to work stations should be provided.

4.2 Stairs and passageways. Stairs (often called "ladders") on ships, where possible, should run fore and aft. If run athwartships, steps can become dangerous whenever the vessel rolls. Passageways should be as straight as possible and wide enough to permit easy passage of furniture. Handrails should be fitted in passageways as well as along stairs. Figure 3.6, which shows a typical arrangement of the crew quarters in a cargo ship, indicates the key role played by the passageways and stairs in the overall arrangement.

Doors deserve careful thought. Like corridors, they should be wide enough to admit furniture. Their location, hinging, and direction of swing will have an influence on accessibility and convenience in living and working aboard.

Further Reading

Tapscott, Robert J., "General Arrangement," Chapter III in *Ship Design and Construction*, Robert Taggart, Ed., Society of Naval Architects and Marine Engineers, Jersey City, N.J., 1980.

Muller, W. H., "Some Notes on the Design of Crew Accommodations for Merchant Vessels," *Transactions*, Society of Naval Architects and Marine Engineers, Vol. 67, Jersey City, N.J., 1959, pp. 715–756.

Benford, Harry, "Ship Manning Trends in Northern Europe: Implications for American Shipowners and Naval Architects," *Transactions*, Society of Naval Architects and Marine Engineers, Vol. 92, Jersey City, N.J., 1984, pp. 303–330.

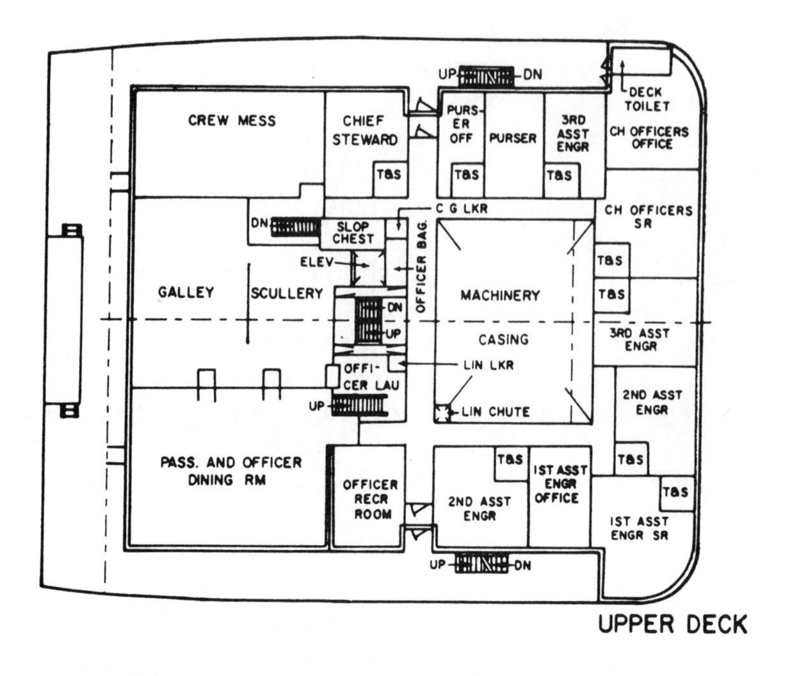

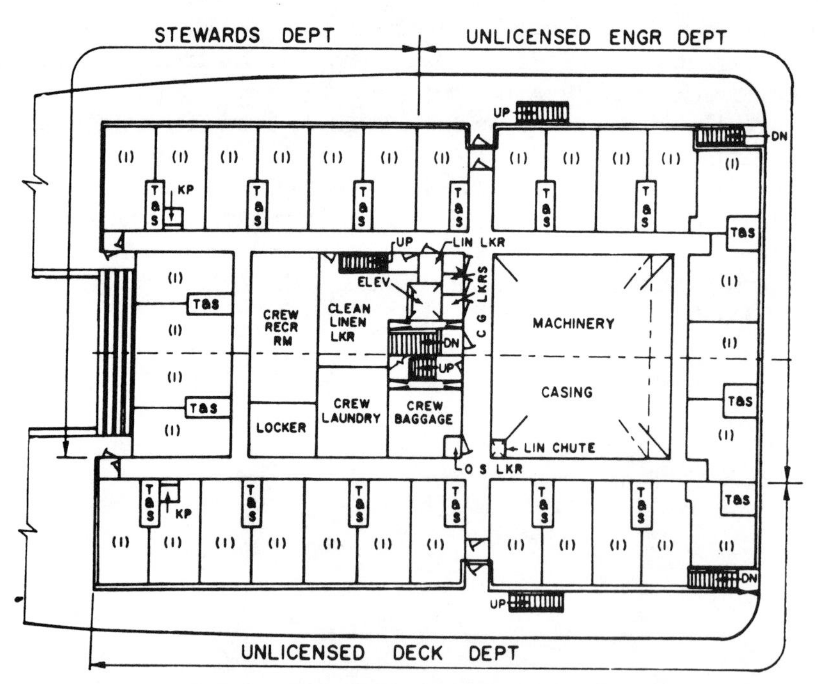

Fig. 3.6 Layout of crew quarters on a cargo ship (source: Tapscott, 1980).

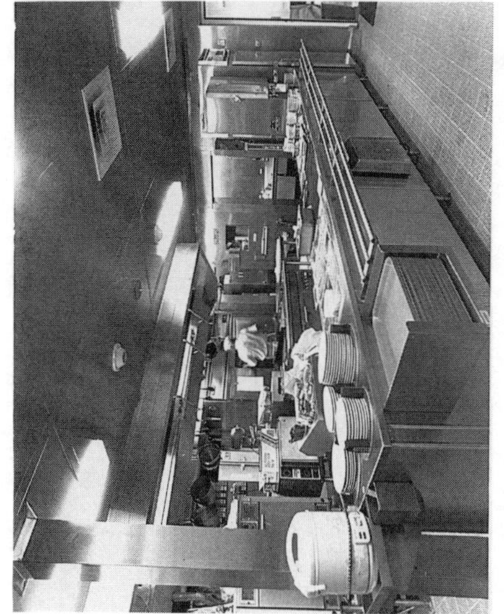

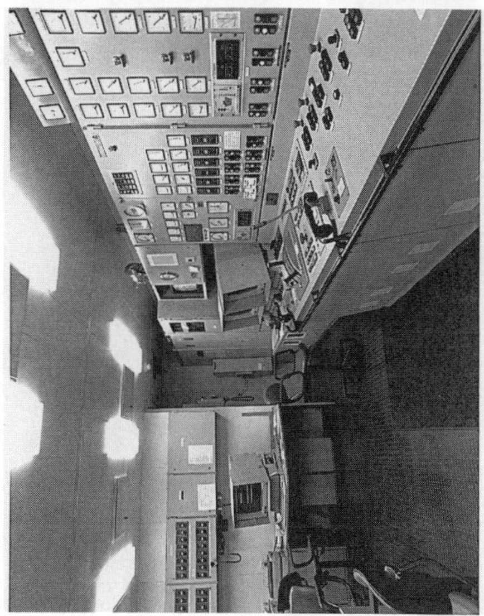

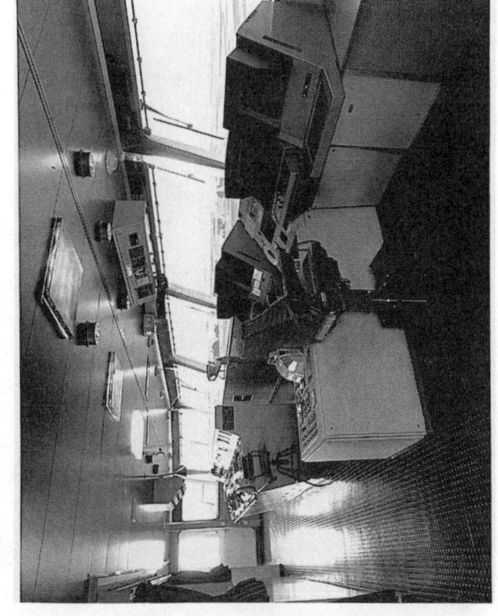

Fig. 3.7 Interiors of a modern containership. Upper left: The galley. Upper right: The crew's mess. Lower right: The bridge, with seat and control equipment for navigating officer at center and communication, monitoring, and steering control in background. Lower left: Machinery monitoring and control center. Courtesy Thomas Winslow, American President Lines, Ltd.

Fig. 3.8 Accommodations in a modern containership. Upper left: Officers' lounge and bar. Upper right: A senior officer's sitting room. Lower left: Crew's exercise and sports room. Lower right: An officer's private cabin. Courtesy Thomas Winslow, American President Lines, Ltd.

CHAPTER IV

DESIGN PROCEDURES

1. Creating a Design

1.1 General perspective. When naval architects talk about "designing" a boat or a ship, two basic phases of the work are often implied: the creative and the analytical. In this chapter, however, I shall concentrate solely on the creative aspects of design.

In the creation of a design, the logical procedure to follow will depend on many combinations of circumstances. One set of circumstances pertains to the client's needs and constraints. Another set pertains to the naval architect's intellectual and technical tools, his or her experience, his or her assistants (if any), the time available to do the work, and so forth. Surely it is safe to say that there are unlimited numbers of logical procedures to be followed in working up designs. The best I can hope to do in this chapter is to present a few sample sets of circumstances and suggest in each case a logical course to follow in producing a design. Do not be misled; I am not trying to give you a detailed how-to-do-it cookbook for ship design. I shall simply outline a few logical procedures.

The simple end of the spectrum is the place to start.

1.2 Just go ahead and build it. The design of some boats exists only in the brain of the builder. Want a dugout canoe? Look for the biggest straight-trunk tree within reasonable dragging distance of the shore. In a more civilized environment you may find rowboats tending to be of length exactly equal to the longest available sheet of marine plywood.

1.3 Modify. Perhaps the most common approach is to copy and possibly improve on the design of some existing, presumably successful, craft whether large or small. This is carried to the extreme when you borrow a friend's canoe and use it as a mold for your own fiberglass clone.

1.4 Dream. A high form of doodling involves the idle creation of designs for boats or ships. These designs may be based on some perceived need, or you may simply let your pencil go into business for itself. There is nothing wrong with daydreams, and they may produce highly original and admirable concepts, but considerably more rigor is appropriate before starting to build.

1.5 Aim to satisfy some functional need. Now we must recognize that the naval architect's central role in life is to satisfy some client. This leads to a logical approach to design based on what is called *systems analysis*. In this procedure engineers attack a problem in the following systematic steps:

a. Define the client's needs in functional terms. For example: Move X tons of commodity Y from Port A to Port B each year.
b. Determine the constraints under which the system must operate. What safety regulations must be met? What labor agreements must be honored? What external conditions place limits on overall dimensions? And so forth.
c. Agree upon a measure of merit. How do we define "best"?
d. Generate an exhaustive list of alternative strategies, each of which will satisfy the client's functional needs in the face of the constraints. We talk about *strategies* here rather than *designs* so as to encompass operating procedures as well as hardware. Each proposed alternative should of course be technically feasible.
e. Estimate the quantitative value of the measure of merit for each of the strategies.

Having completed the work outlined above, the naval architect should go back to the client and discuss the results. Although much of the analysis may have been conducted with the aid of a computer, the responsibility for selecting the actual design should fall to the client in consultation with the naval architect. No one but the client is qualified to weigh the many intangible factors that should influence the decision. By *intangible factors* I refer to the set of important influences whose impact cannot be weighed in the scale of the measure of merit. For example, how important to your client is the outward appearance of the proposed vessel? How do you apply numbers to that question? The design promising the best value of the measure of merit will not necessarily promise the client the most personal satisfaction.

What I have outlined immediately above represents something of an ideal set of circumstances. In real life, few prospective ship or boat owners are likely to specify their needs in purely functional terms. Ask any yacht architect.

This systems analysis approach may be used at many stages during the design process. It is perhaps best suited to the preliminary stage of design, in which such characteristics as the vessel's main dimensions, speed, and power are selected. But this approach to decision-making can also be kept in mind as the vessel's design details are developed. How will the ship's overall measure of merit be affected by each decision? Clearly, as the naval architect moves beyond the early key questions, the client need not be consulted about every decision. There are, however, many clients who will want to follow every step along the way and they are normally accorded that privilege.

1.6 Limits. In the development of a list of alternative strategies the general approach will be influenced by the nature of the limits that may apply. In the design of an America's Cup boat, cost may be relatively unimportant, but for most yacht designs there will be upper limits on what the client is willing to spend.

In merchant ships there are two families of size constraints, depending on the availability of cargo (or passengers, or vehicles). In some bulk cargo trades, such as petroleum or iron ore, the cargo is virtually unlimited in supply. In such trades, the best economies are achieved through making the ship as big as external constraints (harbor depths, canal limits, etc.) allow. In designing those ships, the naval architect starts with the largest possible set of dimensions and derives from them the cargo-carrying capacity. Starting with a specified cargo capacity would be a mistake.

In other trades there are definite limits on the supply of cargo, passengers, or vehicles. If the designer tries to exploit the advantages of scale, the end result would be a ship that would spend months trying to find enough cargo to fill the holds. The logical approach is to start with the client's specified cargo capacity and then seek the ideal set of dimensions and other characteristics that will satisfy that specification. This is a more complex problem than that of the bulk carrier with unlimited cargo, and some organized form of trial and error is usually required. When properly established, the procedure will lead ever closer to a design of the desired functional capability. Some form of computer assistance is usually appropriate.

2. *Measures of Merit*

2.1 Perspective. How do naval architects evaluate the relative merits of competing proposals? As is the case with design procedures, different circumstances dictate different measures. Moreover, for any given set of circumstances, there are honest differences of opinion as to the best way to measure "goodness." In this section I outline some of the different circumstances and suggest logical ways to make wise decisions.

2.2 Role of economics. No matter what kind of boat or ship is being designed, naval architects should base their decisions on making the best use of scarce resources: manpower, materials, and money. There may be a few exceptions to this. If the client wants a racing hydroplane and is firmly determined to win some race regardless of cost, then there is but one measure of merit—speed. In most cases, however, several factors must be weighed before making a decision: relative weights, ease of construction, ease of operation, durability, comfort, and so forth. Engineers try to handle these disparate factors by converting everything, as best they can, to dollar equivalents, the nearest thing there is to a universal unit of measurement. This brings in the techniques of applied economics.

2.3 Engineering economics. Almost all engineering projects require someone to make a capital investment. The aim of engineering economics is to make that investment as rewarding as possible. The logical measure of engineering success, then, must involve dollar signs. In the industrial world this means *profitability.* In government projects it means *maximum public benefit per dollar cost.* In pleasure craft it means *maximum personal satisfaction per dollar cost.*

An important basic concept in applied economics is that near-at-hand dollars are more important than far-off dollars. (This is quite aside from the threat of inflation.) Money does you no good until you spend it on your own personal needs and desires. Most of us would rather have a dollar to spend today than the promise of a dollar some time in the future. That is why we must consider not only how much money changes hands, but also *when.* This requires us to discount future cash flows. I do not intend to get into the mechanics of doing that, but merely mention that standard compound interest relationships are applied.

2.4 Alternative economic criteria. As already mentioned, in private industry the measure of merit should be based on profitability. Opinions differ, however, as to the definition of *profitability.* Some business managers try to predict future net income on an annual basis. They discount each annual amount based on a minimum acceptable interest rate (perhaps their own cost of raising capital). From the total of those discounted amounts they subtract the initial investment. The difference is the *net present value,* commonly abbreviated, and called, NPV. Among alternative investment opportunities, they would tend to favor the one promising the highest NPV, all risks being equal.

Another popular measure of merit within the business community goes under various names including *yield* and *discounted cash flow rate of return* (or DCF). In this case, instead of starting with a minimum acceptable interest rate, they find by trial and error the interest rate that will make the NPV equal to zero. If you were to deposit the initial investment in a bank paying that particular rate of interest, you should then be able to withdraw in the future the same amounts that had been predicted as the project's future net incomes. Obviously, that is a good measure of profitability, and you would want to select the project promising the highest yield.

With government-owned vessels such as naval craft, icebreakers, or oceanographic research ships, there is no income to be weighed. In such cases the naval architect tries to predict the initial investment and the future annual costs. The government, like private industry, must discount future monetary amounts because it, too, must recognize the time-value of money. Adding the sum of the discounted future costs to the initial cost produces what is called the *life-cycle cost.* All other things being equal, the alternative concept offering minimum life-cycle cost should be favored.

An alternative measure of merit for non-income-producing ships is called *average annual cost* (AAC). This is *not* simply the average annual operating cost for fuel, wages, etc. It includes all those expenses and also a uniform increment called the *annual cost of capital recovery*. This increment is found by converting the initial investment into annual amounts that, when discounted, are equivalent in value to the initial amount.

Generally speaking, in most circumstances average annual cost and life-cycle cost will lead to the same design decision.

2.5 A simple case. Let us consider the case of a single-product ship engaged in a single trade. A crude oil carrier operating between the Persian Gulf and Rotterdam would be a typical example. If we are confident of our predictions of future freight rates we could base design decisions on either NPV or yield. But, suppose we are forced to admit that we have no logical way to predict those rates? We could then turn to one of the criteria that are independent of income. One of these, as you may recall, is average annual cost: AAC. But, to be realistic we must recognize that the alternative designs probably promise different transport capabilities. What does low AAC mean if it is tied to low productivity? The answer here is to relate AAC and productivity. Productivity in this case would be defined as the annual transport capacity on the given trade route. The ratio of AAC to tons of cargo moved per year will give us what we call the *required freight rate* (RFR). This is the rate that the shipowner must charge the customer if the owner is to regain the investment at some reasonable level of after-tax profitability. The implication here is that the best ship in any given trade is the one that can offer minimum rates while still producing reasonable profits.

Table 4.1 summarizes the key elements of the measures of merit described above.

2.6 Complex cases. Economic analyses become less and less reliable as consideration is given to ships with multiple functions or unpredictable operating conditions. Take the case of a multipurpose oceanographic research ship. Suppose two alternative designs are under consideration, one being far superior to the other but also more expensive as measured by average annual cost. Subjective judgment must be brought to bear. One approach would be to find some common multiple of the average annual costs. Let us say that we find we could get three of the mediocre ships for the cost (as measured by AAC) of two of the superior ships. We should then ask the decision maker whether it would be better to acquire three of the one kind or two of the other.

2.7 Where first cost rules. In some kinds of craft, operating costs are relatively unimportant. This is often true in pleasure-boat design. The owner is willing and able to pay just so much for the boat, but is little concerned about annual costs for fuel, repairs, and so forth.

Table 4.1 Summary of economic criteria applicable to decision-making in ship design

Criterion	Appropriate for	Necessary Inputs[a]
NPV (net present value)	Commercial ships	Annual revenue, Annual operating costs, Minimum acceptable rate of interest
DCF (discounted cash flow rate of return)	Commercial ships	Annual revenue, Annual operating costs
AAC (average annual cost) and LCC (life-cycle cost)	Nonrevenue-producing ships, Naval craft, Coast Guard, service vessels	Annual operating costs, Interest rate
RFR (required freight rate)	Commercial ships	Annual transport capacity, Annual operating costs, Interest rate

[a] All criteria require estimates of first cost (that is, investment), economic life, and tax rate (if any).

Given that certain level of fixed cost, design decisions should be aimed at providing the owner the maximum amount of satisfaction without exceeding that figure. Let the owner decide what is meant by *satisfaction.*

2.8 Summary. We find in naval architecture that design decisions should be based as much as possible on science (economic analysis), but to some degree art (subjective judgment) will always be required. What I have spelled out above will help you understand in a qualitative way how economics can be used to help make good decisions in ship design.

Further Reading

Kiss, Ronald K., "Mission Analysis and Basic Design," Chapter I in *Ship Design and Construction*, Robert Taggart, Ed., Society of Naval Architects and Marine Engineers, Jersey City, N.J., 1980.

Michel, Walter H., "Mission Impact on Vessel Design," Chapter II in *Ship Design and Construction*, Robert Taggart, Ed., Society of Naval Architects and Marine Engineers, Jersey City, N.J., 1980.

Tapscott, Robert J., "General Arrangement," Chapter III in *Ship Design and Construction*, Robert Taggart, Ed., Society of Naval Architects and Marine Engineers, Jersey City, N.J., 1980.

Hamlin, Cyrus, *Preliminary Design of Boats and Ships*, Cornell Maritime Press, Centreville, Md., 1989.
Buxton, Ian, "Engineering Economics Applied to Ship Design," *Transactions*, Royal Institute of Naval Architects, Vol. 114, London, 1972, pp. 409–428.
Benford, Harry, "Of Dollar Signs and Ship Designs" in *Proceedings*, STAR Alpha, Society of Naval Architects and Marine Engineers, Jersey City, N.J., 1975, pp. 14-1 to 14-30.

CHAPTER V

HULL FORMS

1. *General Considerations*

A well-designed hull form is a successful amalgamation of many of the conflicting considerations mentioned in Chapter I. The craft must be beamy enough to be stable, yet narrow enough to be easily propelled. A long narrow ship promises a relatively low-power (hence inexpensive) engine, but an expensive hull; a short, stubby ship promises the opposite—an inexpensive hull, but an expensive engine. A ship that is cheap to build because it is box-like will be expensive to operate because it will have high fuel bills. And so it goes. Balancing these conflicting requirements is a highly subjective art, and no two naval architects will ever design exactly the same hull form.

Success in communication requires a shared understanding of the precise meaning of words. Within the marine community, a basic set of definitions allows a clear dialogue regarding factual, quantitative information about a ship's major dimensions and hull form characteristics. There are, inevitably, at least a few linguistic divergences between the three major sectors of the profession: merchant ships, naval craft, and yachts. Merchant ship definitions are perhaps the most widely used. They apply also to many miscellaneous kinds of vessels such as oceanographic research ships and commercial fishing boats. This chapter, then, stresses commercial ship practice but includes appended comments on variations in definition that exist among designers of yachts or naval craft.

2. *Hull Form Definitions*

2.1 Linear measurements. First, what do naval architects mean by *length*? Figure 5.1 shows the three most important definitions. *Length overall* (*LOA*) is what matters to civil engineers who design locks or wharves. *Waterline length* (*LWL*) is important to hydrodynamicists. (Those are the naval architects who are concerned with a ship's speed and power relationship as well as other motions in the water.) *Length between perpendiculars* (*LBP*) is the dimension used by most other naval architects, including those employed by classification societies. In the U.S. Navy, *LBP* is arbitrarily defined as being equal to

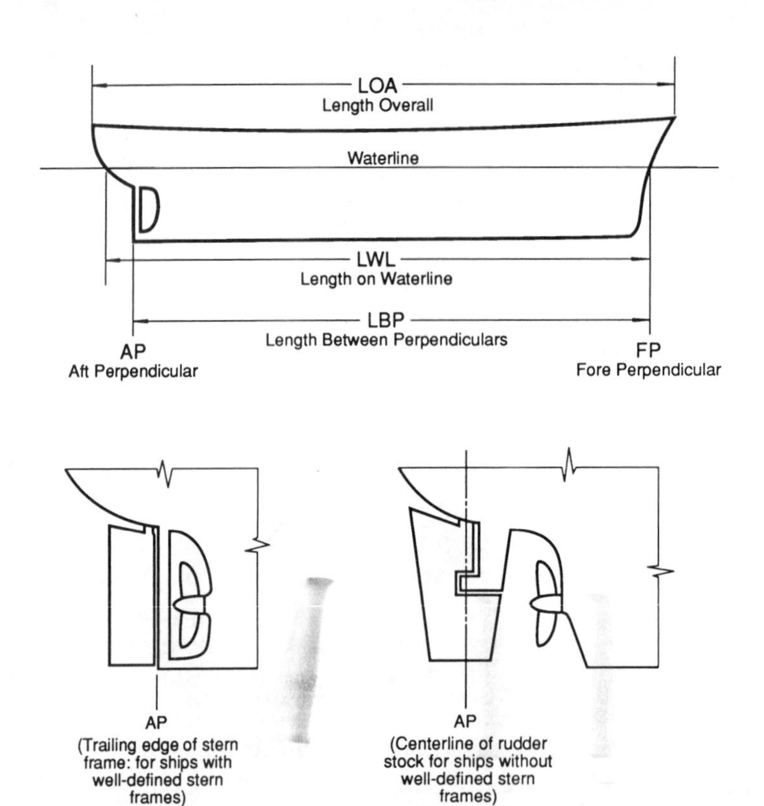

Fig. 5.1 Some standard definitions of length. In merchant ships the waterline would be at the summer saltwater loadline; in others it could be the design loadline.

LWL. In yachting circles, *length* is usually taken as *LOA*, or the length on deck.

Locating the after perpendicular (AP) is tricky. If the ship has a well-defined stern frame, then the after perpendicular will be at its trailing edge, that is, just forward of the rudder. If there is no well-defined stern frame, then the after perpendicular will be at the centerline of the rudderstock (that being the vertical shaft that turns the rudder). (See Fig. 5.1.)

Figure 5.2 defines several terms related to a vessel's transverse shape. The shape shown is that of the *molded hull form*. By this is usually meant the form *inside* the shell plating. (By using the internal dimensions, naval architects simultaneously define the shape of both the plating and the internal stiffening members, called *framing*.) In wooden, or fiberglass, hulls the molded surface is more likely to be the exterior surface.

The symbol ⊞ means *baseline*, while the symbol ℄ means *centerline*. (Actually, since these two key reference lines each show up in each of two views, they really represent planes.) The *molded half beam* ($B/2$) is taken at the ship's maximum molded width. *Deadrise* (or *rise of floor*) may be incorporated in order to make it easier to pump out liquids from within the ship's bottom. A ship with deadrise will also usually have *half siding* so as to prevent mutual damage between hull and keel blocks when in dry dock. (If you need help with those terms, please see the Glossary.) The purpose of the *bilge radius* is to ease the flow of water from bottom to sides when the ship is sailing in trim, for example, with the stern deeper in the water than the bow.

When a ship is scraping along a lock wall, or tied up alongside another ship, *tumblehome* spreads the area of contact and so saves the upper corner of the hull from concentrated abrasive damage. At the bow of the ship, the sides may sweep out in a concave form called *flare*, which has no well-defined measure. Flare has two purposes: it increases the deck area in the bow (making more space for mooring gear) and tends to deflect waves, and so leads to a drier deck.

Camber (sometimes called *round of beam*) helps the weather (that is, uppermost, exposed) deck shed water. Its form may be curved or made up of two or more flat, sloped surfaces. The amount of camber is generally taken as a fraction of the beam (perhaps 2 percent). For this reason, its measure is taken (as shown in Fig. 5.2) from the theoret-

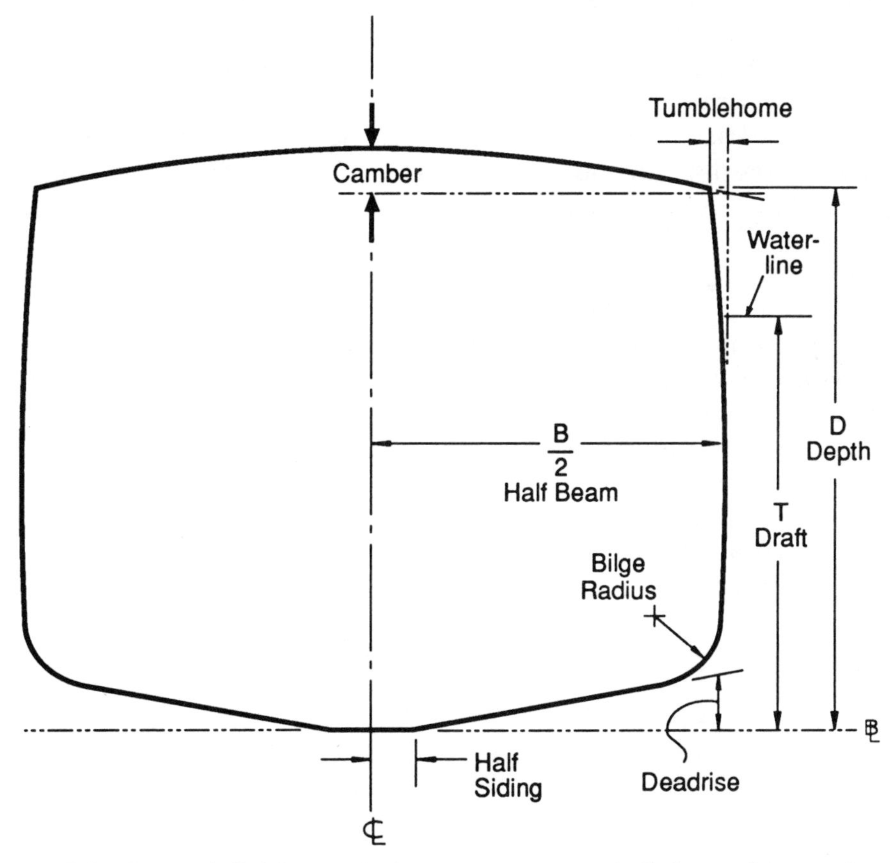

Fig. 5.2 Some definitions relating to transverse hull form characteristics. Freeboard is approximately equal to depth minus draft.

ical intersection of the extended line of the camber curve and the vertical extension of the molded half-beam line.

Freeboard refers to the vertical extent of the hull that is out of the water at any given time. In merchant ships there are internationally agreed-upon rules that establish safe minimum values. The basic measure of freeboard assumes the ship to be floating in salt water and to be operating during the summer (that is, most favorable) season. This is called the *summer load line* and is identified on both sides of the ship at mid-length by what are usually known as the *Plimsoll marks*. Auxiliary marks show safe freeboards under other conditions, such as winter operations, or when floating in fresh water. A practical peculiarity of the freeboard rules is that the measure is taken down from the *top* of the deck, rather than from the molded surface.

Following Archimedian principles, a floating vessel will sink into the water until it has pushed aside its own total weight of water. *Draft* (*T*), then, will be a function of the extent of variable weights placed aboard. In short, we must be careful in defining *draft*. In merchant ships the standard value is taken at the summer load line. If the ship is intended to carry light, voluminous cargoes it may normally sail with the Plimsoll marks high and dry. In that case the hull form may be designed around a greater freeboard, leading to the concept of a design draft that is significantly less than the summer loadline draft.

The *keel draft* is measured up from the lowest point on the outside of the hull. If an underhung keel is fitted, the keel draft may be substantially greater than the molded draft. Those painted draft numerals that you may have seen on bow and stern are based on the keel draft.

The *depth* (*D*) has already been defined in the first chapter; but be careful to note that it is measured up to the underside of the deck at edge, not at the centerline.

The depth of water in many harbors and canals severely limits the operating draft of ships. Therefore, in order to maximize carrying capacity, the ships are usually designed to operate on an even keel when in the full load condition. In other words, they will have zero *trim*. Many smaller craft, however, are designed with the stern lower in the water than the bow. A boat in that condition is described as having a *trim aft*, or *keel drag*. When that is the intent, the baseline is usually drawn parallel to the design load waterline and at a height

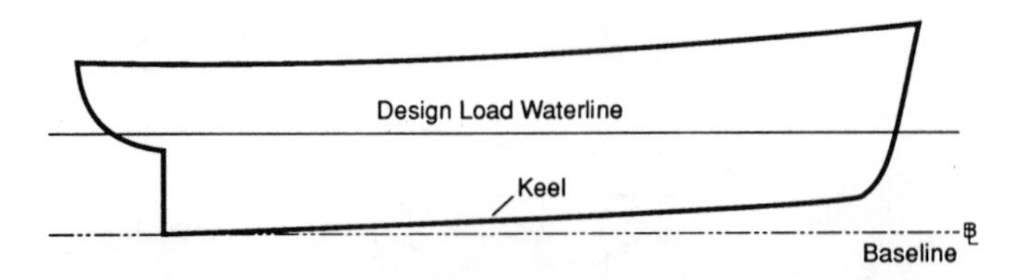

Fig. 5.3 Standard location of baseline in a ship designed with a trimmed keel line. Baseline is parallel to design load waterline.

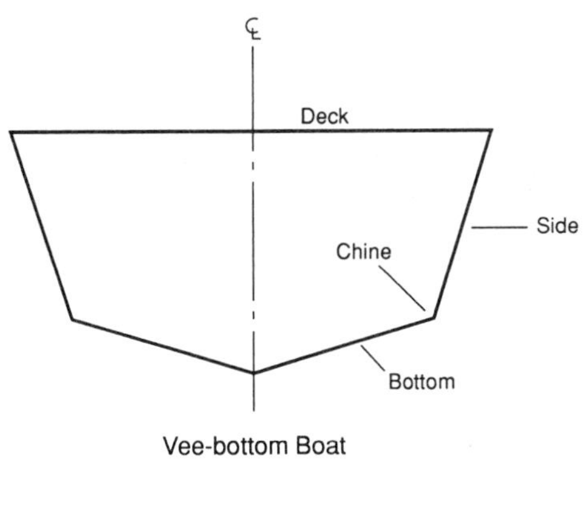

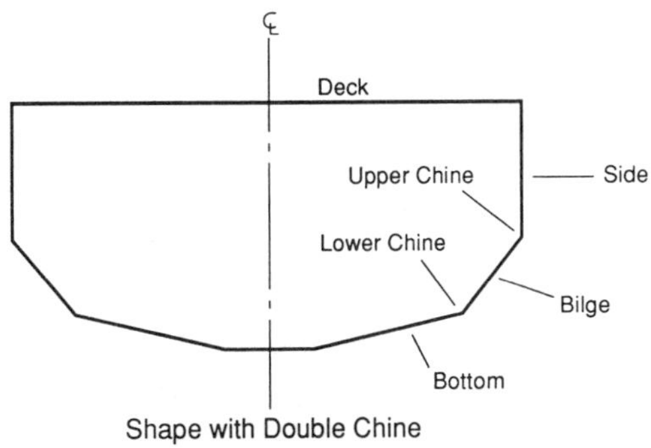

Fig. 5.4 Showing how chine lines may be used in place of bilge radius.

coincident with the lowest point of the molded hull form (see Fig. 5.3). There are variations on this and some yacht architects draw the baseline at an arbitrary distance below the molded hull form.

2.2 Two- and three-dimensional characteristics. *Knuckles* are concave or convex joints in plating. Where one or more fore-and-aft knuckles are used in place of a bilge radius, they are called *chines* (see Fig. 5.4). In planing craft, V-shaped bottoms and chines are necessary. (A *planing craft* is one that, when driven at high enough speed, rides up on the water surface supported primarily by the impact of the water on its bottom. It is in contrast to the displacement-type hull.) Knuckles or chines may be used to save money by making ships easier to build. The V-bottom boat shown later in Fig. 5.9 has what is called a *developable surface*. This means that the hull can easily be formed of materials, such as plywood, that are otherwise not readily bent into three-dimensional curves.

On some occasions, naval architects find it important to differentiate between *molded displacement* and *total displacement*. As you may recall, the molded surface is usually defined as being inside the shell plating. The molded displacement, then, would be based on that imaginary interior volume of the hull form below any given waterline. If we hold that same waterline, the *total* volume, hence weight, of displaced water would be slightly greater. There are two components of the increase. The first is the thickness of the shell plating (which in a big ship may add a half percent to the molded full-load displacement). The second is the volume of what we call the *appendages*. These are exterior fittings such as rudders, propeller shafts, or bar keels.

Underwater volume (V) and corresponding displacement (Δ) are related by the density of the liquid in which the ship is floating. The density of fresh water is such that one long ton of it occupies 35.9 cubic feet. The density of salt water, on the other hand, is such that one long ton occupies 35 cubic feet. This means that, when using British units, the displacement in long tons may be found by dividing the ship's underwater volume (in cubic feet) by 35.9 if in fresh water, or by 35 if in salt water. In SI units, the freshwater displacement in tonnes will exactly equal the volume in cubic meters. (That is, indeed, how tonnes are defined.) The salt water displacement will be 2.5 percent greater.

Sheer is the upward curvature of the deck as viewed from the side. If the deck has camber, there will be two sheer lines: the deck at edge and the deck at centerline. (If the deck is without camber, the two sheer lines will coincide.) The low point of sheer is usually at mid-length, and the sheer at the bow will typically be twice the sheer at the stern. Sheer curves are often parabolic, but the designer is free to choose whatever shape pleases the eye.

2.3 Hull form coefficients. We must next consider a family of dimensionless coefficients that tell us something about the character of the underwater hull form. The first of these is the *block coefficient* (C_B). It is a measure of the "blockiness" or fullness of the underwater hull. Its numerical value equals the ratio of the ship's molded volume

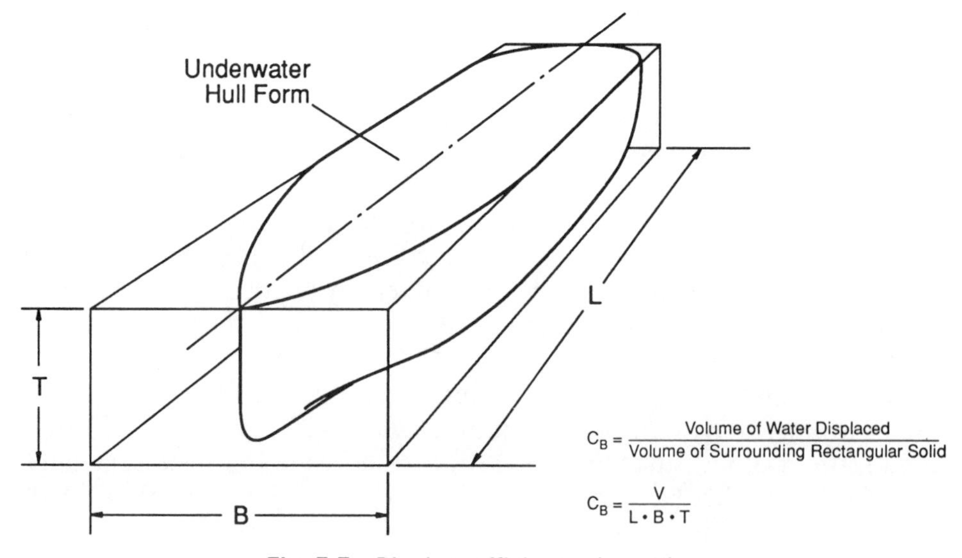

Fig. 5.5 Block coefficient schematic.

(below the design waterline) to the volume of the surrounding (imaginary) rectangular solid (Fig. 5.5). Typical values of C_B range all the way from 0.50 for an oceanographic research vessel or tugboat to 0.90 for a Great Lakes bulk carrier. For a rectangular pontoon the value would of course come out to unity. In equation form we have:

$$C_B = \frac{V}{L \cdot B \cdot T}$$

where

V = volume of displacement (molded)
L = length (either *LWL* or *LBP*)
B = beam
T = draft

Note that length may be either *LWL* or *LBP*. Either is acceptable as long as you are consistent. High values of the block coefficient are associated with relatively low-speed vessels that are affected by exterior constraints (such as locks or harbors) on their overall dimensions. To maximize displacement, hence cargo capacity, the block coefficient must be made large. Low values, on the other hand, are appropriate where exterior dimensions are not particularly constrained. That being the case, low values of block coefficient lead to low hull form resistance and good seakeeping qualities.

The *maximum section coefficient* (C_x) is taken at that place along the ship's length where the transverse underwater area is greatest. Its numerical value equals the maximum area (A_x) divided by the area of the surrounding rectangle, beam times draft:

$$C_x = \frac{A_x}{B \cdot T}$$

A closely related measure is the *midship coefficient* ($C_{\otimes}$). In this variation, the transverse area at the ship's mid-length ($A_{\otimes}$) is used in place of A_x. In most cases the values coincide and there is no difference in value between $A_{\otimes}$ and A_x. In the rare cases where that is not true, one is better off using A_x.

A measure of particular interest to hydrodynamicists is the *longitudinal prismatic coefficient* (C_p), usually called simply the *prismatic coefficient*. It is the ratio of the displaced volume of water to the volume of an imaginary prism with a base area of A_x and a length of *LBP* or *LWL* (Fig. 5.6). In equation form we have

$$C_p = \frac{V}{L \cdot A_x}$$

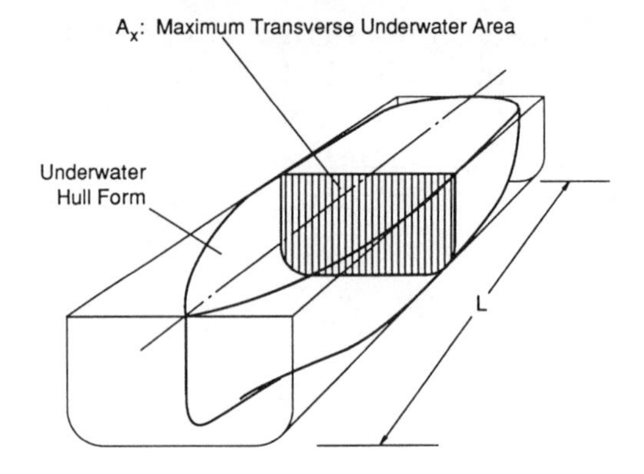

$$C_P = \frac{\text{Volume of Water Displaced}}{\text{Volume of Surrounding Prism}}$$

$$C_P = \frac{V}{L \cdot A_x}$$

Fig. 5.6 Prismatic coefficient schematic.

If you will play simple algebraic games with the equations for C_B, C_x, and C_p, you can prove to yourself that

$$C_p = \frac{C_B}{C_x}$$

Fig. 5.7 illustrates a fourth important value, that of the *waterplane coefficient* (C_{wp}): the ratio of the area of the waterplane (at the design draft unless otherwise stated) to the area of the imaginary surrounding rectangle ($L \cdot B$):

$$C_{wp} = \frac{A_{wp}}{L \cdot B}$$

where A_{wp} is the area of the waterplane.

Now, although the foregoing coefficients can tell us a good deal about the nature of hull forms, they tell us nothing about whether the form is short and fat or long and slim. Moving a boat sideways instead of forward will not change its block coefficient, but it will certainly change its resistance to moving through the water. We therefore need nondimensional measures of fatness versus slimness when trying to predict speed and power relationships. These can take any of the following forms:

$$\frac{V}{L^3} \quad \text{or} \quad \frac{L^3}{V} \quad \text{or} \quad \frac{L}{V^{\frac{1}{3}}}$$

where L is length and V the volume of displaced water.

Naval architects who use British units may rely on what is called the

$$\text{displacement-length ratio:} \quad \frac{\Delta}{\left(\frac{L}{100}\right)^3}$$

where Δ is in long tons and L is in feet. Obviously, it is dimensionally dependent. That is, its numerical value will change if SI units are used or if fresh water is involved in place of salt water.

Table 5.1 gives typical numerical values of three coefficients for some widely differing kinds of hulls.

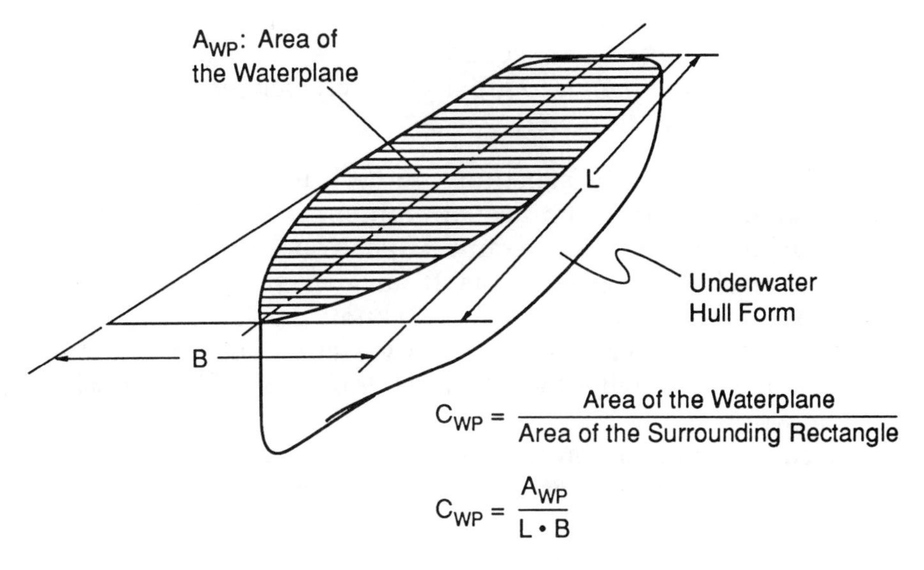

Fig. 5.7 Waterplane coefficient schematic.

2.4 Relative speed. Another important coefficient relates a ship's speed through the water to its size (as measured by length). This concept dates back only to 1868, with William Froude's theories pertaining to ship versus model speed and power characteristics. You will learn about Froude's theories in Chapter VIII. For now I shall merely define his nondimensional speed-size ratio, the so-called *Froude number* (FN):

Table 5.1 Representative hull form coefficients

	Type of Hull		
	Oceanographic Research Vessel	Power Yacht	Great Lakes Bulk Carrier
General Characteristic	stubby and rounded	slim and fine-lined	slim and box-like
C_B	0.58	0.56	0.87
V/L^3	0.0152	0.00566	0.00281
$\Delta/\left(\frac{L}{100}\right)^3$	433	162	80

$$\text{FN} = \frac{v}{\sqrt{g \cdot L}}$$

where

v = speed in meters per second
g = acceleration owing to gravity: meters per second per second
L = ship's length in meters

If using British units, substitute numerical values appropriate to feet in the above. Since the Froude number is nondimensional, the numerical outcome will be the same. High Froude numbers are indicative of relatively high speed. A 500-ft naval craft moving at 35 knots would have a value of about 0.45. At 10 knots the value would drop to 0.13. A 150-ft oceanographic research vessel moving at the same 10 knots would have a higher Froude number because of its lesser length. Its value works out to be 0.24.

As a cultural matter you should know that Mr. Froude pronounced his name to rhyme with *brood,* not *crowd,* nor with that Viennese psychiatrist.

Although the Froude number is popular with hydrodynamicists, other naval architects are more likely to substitute the *speed-length ratio,* which is found by dividing the speed in knots by the square root of the length (usually *LWL*) in feet. As you will note, this ratio is simpler than the Froude number because it uses the more common unit of knots in place of feet per second, and drops g because the force of gravity remains essentially constant over the ocean's surface. It is, however, dimensionally dependent and will produce a different number if SI units are used.

Here is how you can convert the speed-length ratio to the corresponding Froude number:

$$FN = 0.298 \frac{V_k}{\sqrt{L}} \quad \text{or, say} \quad \frac{3}{10} \cdot \frac{V_k}{\sqrt{L}}$$

where V_k is speed in knots, and L the waterline length in feet.

2.5 Hull proportions. There are many other useful coefficients that quantify hull form characteristics. Among these are four simple two-part dimensional ratios:

- the *length-depth ratio*, useful in structural design,
- the *length-beam ratio*, a rough measure of leanness with implications about maneuverability,
- the *length-draft ratio*, which is indicative of likelihood of lower-bow slamming damage during severe weather,
- the *beam-draft ratio*, with implications for transverse stability and wave-making characteristics.

In understanding the nature of how a buoyant body will float and remain upright, naval architects must be able to locate the center of volume of the displaced water. This is commonly called the *center of buoyancy*, and one needs to know both its vertical and fore-and-aft locations. (Since most ships are symmetric port and starboard one can assume the lateral center will be on the ship's centerline when floating upright.) The *vertical center of buoyancy* (VCB) is measured up from the baseline. The *longitudinal center of buoyancy* (LCB) may be measured from either the fore-or-aft perpendiculars. More often, however, it is measured from the midpoint of the perpendiculars, indicated by the *midship mark*: (⌧).

Not to be confused with LCB is the *longitudinal center of flotation* (LCF). That is the fore-and-aft location of the center of area of the waterplane at which the ship is floating. If you want to add a weight without changing the trim, place it over LCF, *not* LCB. A useful thing to know.

You may sometimes hear reference to a ship's *wetted surface*. That is simply the area of the underwater hull and appendages, measured in square feet or square meters. The amount of wetted surface is an important factor in estimating the resistance to motion through the water. It is also used in estimating required amounts of bottom paint.

The *cubic number* (CN) is a rough measure of the overall size of a hull. In early design stages it is used to estimate the ship's internal volumetric capacity as well as steel weight and cost. In equation form we have

$$CN = \frac{L \cdot B \cdot D}{100}$$

where

L = length between perpendiculars
B = beam
D = depth to uppermost continuous deck

3. *Delineating the Hull Form: The Lines Drawing*

3.1 General approach. The naval architect who wishes to define the exact shape of a ship has the problem of depicting a curved three-dimensional surface on a flat sheet of paper. The solution to this task is to use contours to show the two-dimensional shape at several imaginary parallel crosscuts. That is exactly the same approach used by cartographers in their contour maps (showing a series of curving lines each of constant elevation).

If you were to make imaginary cuts along a ship's length, as in slicing a loaf of bread, you could show their shapes in what naval architects call a *body plan* (see Figs. 5.8 and 5.9). Since ships are usually symmetric port and starboard, only one side need be shown. The forebody lines are by convention shown to the right, the afterbody lines to the left, in the body plan. The imaginary crosscut planes are called *stations*. They are numbered from bow to stern in this country, but frequently the opposite way abroad. The initial station is always identified as number zero, *not* one.

If curves showing the profile shapes of bow and stern are added, they and the body plan would give a complete definition of the hull form. For most calculations of stability, resistance, and so forth, that would be all that would be needed. For various other reasons, however, naval architects almost always find it desirable to present two other families of contours. One of these shows the shapes of imaginary planes cutting through the molded hull form parallel to the base plane. These are called *waterplanes* or *waterlines*. Their shapes, as viewed from above, or below, are shown on what is called the *half-breadth plan*. Again, because of symmetry, only one side need be drawn. These are the lines most closely akin to the cartographer's contour lines (see Figs. 5.8 and 5.9).

The third series of contours shows shapes made by imaginary vertical cuts in parallel planes running from bow to stern. They are called *buttock lines* and are shown on the *sheer plan*, sometimes called the *profile* (again, see Figs. 5.8 and 5.9).

There may be additional, auxiliary cuts producing one or more curves such as the *bilge diagonal* shown in Fig. 5.8. The location of such a diagonal is shown in the two sloping lines identified as *WZ* on the body plan. The curved line that you see below the half-breadth plan shows the shape of the imaginary cut when viewed at right angles to the cutting plane.

As you will note in the figure, the locations of the stations are shown in two views (waterplanes and sheer plan) while the shapes of the stations are shown in but one view: the body plan. The same general idea applies to the waterlines and buttock lines: Their locations show in two views, their shapes in but one.

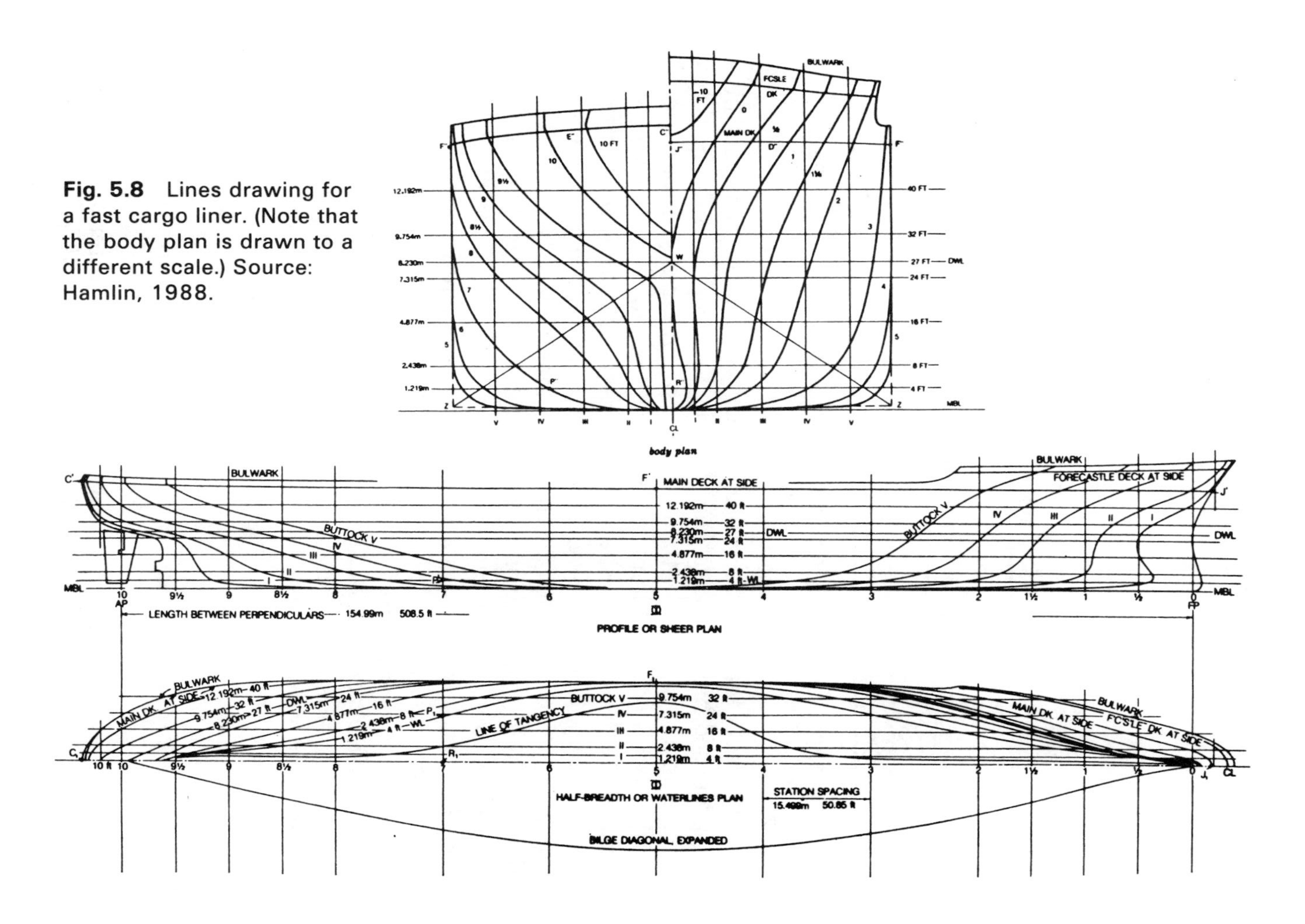

Fig. 5.8 Lines drawing for a fast cargo liner. (Note that the body plan is drawn to a different scale.) Source: Hamlin, 1988.

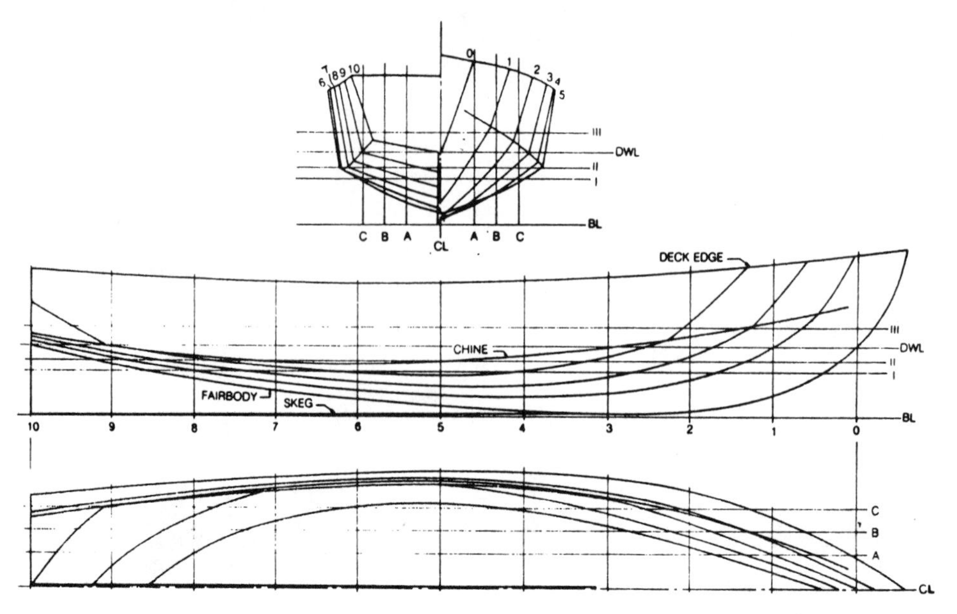

Fig. 5.9 Lines drawing for a V-bottom boat (source: Hamlin, 1988).

3.2 Techniques. A naval architect wants to be sure that his or her proposed hull shape is *fair*. That term pertains to the quality of being smoothly curved, with no abrupt changes in degree of curvature (aside from a reasonable number of well-placed chines or knuckles). A fair three-dimensional shape will be assured if all three families of contours look fair. Drawing fair curves is easy enough. What is difficult is assuring that each of the three major views shows the same identical hull form. This is done through a laborious, trial-and-error process of adjusting the individual lines until their intersections with the various reference lines are in exact agreement in all three views. Explaining this in a book is close to impossible. But, since it is not my intent to make you into a naval architect, you really need not know all the difficult details. Be thankful.

Naval architects are making steady progress in using computers in the lines-fairing operation. The day will come when most of the drudgery will be eliminated.

4. *Hydrostatic Curves*

From day to day a ship may be loaded to different drafts and different trims. There is accordingly a frequent need for some way to find the underwater hull form characteristics over a range of loading conditions. This is done by calculating each characteristic at each of several arbitrary waterlines. These waterlines are always taken parallel to the baseline, but adjustment parameters are included that will allow a properly trained person to make corrections if the vessel is trimmed by bow or stern. The results of these calculations are plotted on closely spaced grid paper, and curves are then faired through the plotted points. These curves are called, collectively, the *hydrostatic curves* or *curves of form*. Figure 5.10 shows such a set in greatly simplified form. Note that the vertical scale shows the ship's draft. A complete set of curves, drawn to various scales, would show:

Displacement (salt water and fresh water, molded and total)
VCB: vertical center of buoyancy
LCB: longitudinal center of buoyancy
LCF: longitudinal center of flotation
C_B: block coefficient
C_x: maximum section coefficient
C_p: prismatic coefficient
WS: wetted surface
KM: location of transverse metacenter above the baseline (explained in Chapter VI)
Moment to trim one inch (or one centimeter)
Change in displacement for one inch (or one centimeter) trim
Tons per inch (or centimeter) immersion

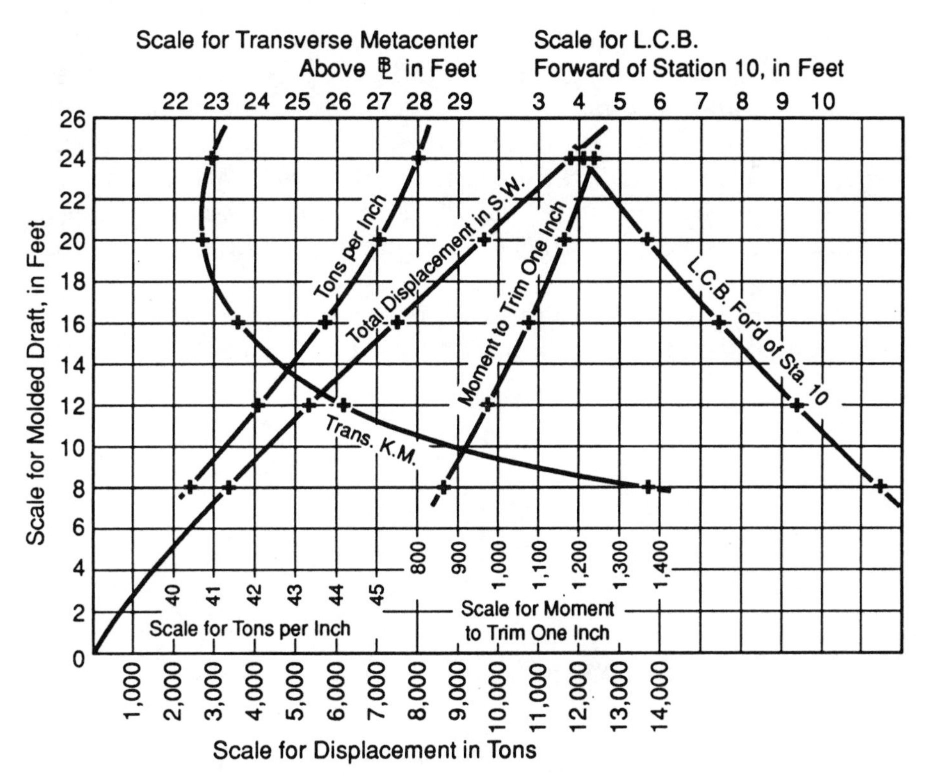

Fig. 5.10 Sample hydrostatic curves.

For the convenience of the deck officers, much of the numerical information shown on the hydrostatic curves is repeated in the form of tables, which most people find easier to use. In cargo ships this information is incorporated in the *capacity plan*, which also shows the volume of each hold and tank and its center of gravity. With that information at hand, the officers can predict the ship's drafts, fore and aft, and stability characteristics for any proposed condition of loading.

5. Appendix to Chapter V: Mensuration

5.1 Preface. The remaining material in this chapter goes somewhat beyond the appropriate depth of coverage for this book's intended readers. I include it for the benefit of those who wish to dig just a little deeper. They will learn how naval architects attack the problem of finding the hull form characteristics (such as those shown on the hydrostatic curves). The task is complex because there is no equation that will define the three-dimensional shape of the typical hull form. The techniques that are employed to overcome this problem may be considered to be the very essence of the science of naval architecture.

5.2 Technique. Naval architects overcome the problems mentioned above by first measuring, and recording, *many* individual dimensions that serve to define the hull form. They then manipulate these measurements in complex, methodical procedures that lead to the desired values. The principles are relatively simple, but the work was extremely arduous until the advent of computers. The generic term for these methods is *numerical analysis*. In naval architectural practice, the great majority of such work relies on a variety of numerical analysis called *Simpson's rule* (more precisely, *Simpson's first rule*). Here is how it works.

If the curve B-D-F shown in Fig. 5.11 is a second-order parabola, it can be shown that the area bounded by A-B-F-E will exactly equal

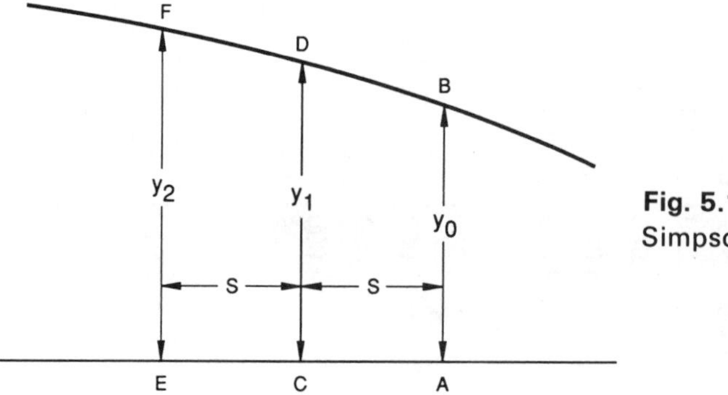

Fig. 5.11 Basis of Simpson's rule.

$$\frac{S}{3}(y_0 + 4y_1 + y_2)$$

If you should take any typical curve (whether station, waterline, or whatever) from a ship's lines, and divide it into enough equal parts along its length, you could safely assume the curve to be reasonably close to a second-order parabola over any two adjacent divisions. (In doing this you must be careful to use an *even number* of equal spaces.) Then, applying the basic method shown above for two spaces to the entire curve, you would find yourself entering the equivalent of the end-y's twice (except at the first and last dividers). The general equation would then become

$$\text{Area} = \frac{S}{3}(y_0 + 4y_1 + 2y_2 + 4y_3 + 2y_4, \text{ etc.})$$

where S is the station spacing.

5.3 Example. Now, let me illustrate this idea with a simple example. Suppose I want to find the area of the waterplane whose shape is shown in Fig. 5.12. Since the form is symmetric I need analyze only one side, being careful to multiply by two before I finish (and, believe me, that's easy to forget). My first task is to divide the baseline—which in this case is the ship's centerline—into some number of equal spaces. In real life these would typically number twenty, but I shall use only six here in order to simplify the explanation. As you may recall from our discussion of the lines drawing (Section 3), these fore-and-aft dividers are called *stations*. In the figure I have identified them with numbers 0 to 6. I am now ready to put Mr. Simpson to work. Using an appropriate scale, I measure the full-size distances from the centerline to the curve at each station. In general terms these distances are called *offsets*. In this particular case they are called *half-breadths*. These distances are shown in the second column of Table 5.2 just below the figure.

The third column in the table, identified as "SM," shows Simpson's multipliers (the 1, 4, 2, 4 , etc., numbers explained above). The final column shows the product of the half-breadth measurements and Simpson's multipliers. The sum of all those products, when multiplied by two-thirds the station spacing, will yield a close approximation to the waterplane area—which is what I set out to find.

I have just explained how to apply the principles of numerical analysis to approximate the area of a waterplane. Naval architects use exactly the same procedure to find the area of any station below the design waterline. That is, instead of analyzing a horizontal area they analyze a vertical area. They do this for several stations along the vessel's length. These cross-sectional areas are then plotted to some convenient vertical scale against their fore-and-aft location, as shown in Fig. 5.13. The smooth line that you see drawn through those data points is what naval architects call the *sectional area curve*, an important tool in ship design.

In this case I have derived a sectional area curve from a set of lines. If I now apply Simpson's rule to that curve's offsets, I can derive the ship's volume of displacement and its longitudinal center of buoyancy. In actual practice, naval architects often work in the opposite direction. That is, they start by drawing what they know to be a good sectional area curve and use that to develop the individual stations, and then fair up the complete lines drawing.

This brings up the question of what is meant by "a good sectional area curve"? It is one that will provide the required displacement with a longitudinal center of buoyancy that will lead to minimum

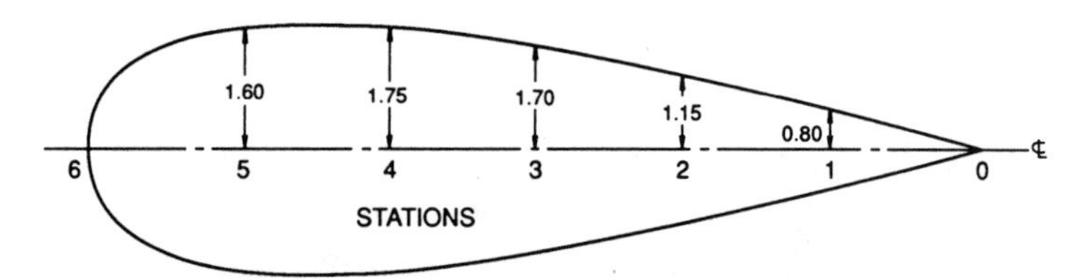

Fig. 5.12 Shape of waterplane, with half-breadths shown in meters (stations are spaced 1.6 m).

Table 5.2 Example showing use of Simpson's rule[a]

Station	Half-Breadth, m	SM	Product
0	0	1	0
1	0.80	4	3.20
2	1.15	2	2.30
3	1.70	4	6.80
4	1.75	2	3.50
5	1.60	4	6.40
6	0	1	0
			Σ Product = 22.20[b]

Area, one side of centerline = (station spacing ÷ 3) · Σ Product

$$\text{Area, both sides} = \frac{2}{3} S \, (\Sigma \text{ Product})$$

Noting that in this case $S = 1.6$ m, we have

$$\text{Area} = \frac{2}{3} \times 1.6 \times 22.20 = 23.7 \text{ cubic meters}$$

[a] Objective: Find area of waterplane shown in Fig. 5.12.
[b] Σ Product is shorthand for sum of the products.

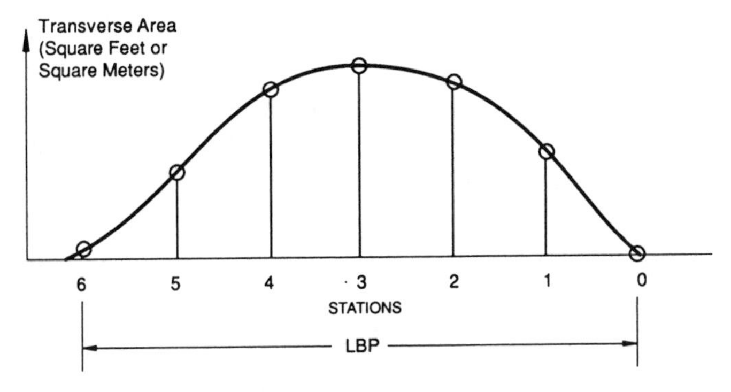

Fig. 5.13 Sectional area curve. The area under the curve (applying full-scale units) represents the volume of water displaced by the hull.

wavemaking resistance (Chapter VIII). It will also result in acceptable trim fore and aft when the ship is in full-load condition. Hamlin (1988) explains how to generate what experience has shown to be sectional area curves of close-to-optimal shape.

This chapter has given you an introduction to numerical analysis in naval architecture. You will probably run across many related matters that may arouse your curiosity: half stations, discontinuities in curves, use of the planimeter, and lots more. The readings cited below will answer your questions.

Further Reading

Barnaby, Kenneth C., *Basic Naval Architecture,* Hutchinson's Scientific and Technical Publications, London, 1948.

Baxter, Brian, *Naval Architecture,* Teach Yourself Books, Holder and Stoughton, London, 1976.

Hamlin, Norman, "Ship Geometry," Chapter 1 in *Principles of Naval Architecture,* Edward V. Lewis, Ed., Society of Naval Architects and Marine Engineers, Jersey City, N.J., 1988.

Rawson, K. J., and Tupper, E. C., *Basic Ship Theory,* Vol. 1, Longman, London, 1976.

Robb, Andrew McC., *Theory of Naval Architecture,* Charles Griffin & Company, London, 1952.

Fig. 5.14 Launch of tanker *Potomac Trader*. This ship is intended to carry several grades of petroleum products. As with most bulk carriers, economic factors dictate a relatively low speed and rather full hull form (that is, high block coefficent). Some indigenous artist could not resist adding that final happy touch on the bulbous bow. Courtesy Peter E. Jaquith, National Steel and Shipbuilding Company.

Fig. 5.15 The fine lines of a cruise ship. The *Sovereign of the Seas* while still under construction. Note the contrast in fullness of form with that shown in the previous figure. The extreme overhang of the stem is largely for visual appeal, but the bulbous forefoot is there to reduce wave-making resistance. Note the two side-thruster tunnels at after end of bulb. Courtesy Royal Caribbean Cruise Line.

CHAPTER VI

STATIC STABILITY

1. Introduction

When we say a boat is *stable* we mean it will (a) float upright when at rest in still water and (b) return to its initial upright position if given a slight, temporary deflection to either side (that is, heeled) by some external force. We then say it has *positive stability*. An *unstable* craft, on the other hand, if slightly deflected, will continue to heel and finally come to rest at some other position. In the worst case it will capsize. Such a vessel is said to have *negative stability*. In the rare case of *neutral stability*, if heeled to some angle by an external force that is then removed, the vessel will remain at the same angle of heel.

Figure 6.1 shows how a cone can be stable, unstable, or neutrally stable depending on its initial position.

This chapter will be confined to the special case of ships at rest in still water. Questions of motions resulting from waves are taken up in Chapter VII.

In the greater part of this chapter I shall try to explain how naval architects analyze the question of whether a proposed ship (when floating in still water) is likely to capsize. This requires an understanding of what is called *transverse static stability*. I shall also consider *longitudinal static stability* (that is, relative to trimming fore and aft) and the special case of submerged submarines. Questions of sailboat stability are touched on in Chapter XI.

I shall explain first the case of intact, undamaged hulls and then present a quick look at what may happen if the shell is ruptured and flooding results.

2. Transverse Static Stability

2.1 The metacenter. Except in heavily ballasted, deep-keel sailboats, the typical vessel's center of gravity is higher than its center of buoyancy. That is, its center of weight (pushing down) is above its center of support (pushing up). You might expect such a combination to be unstable. That is not necessarily the case, however. In a floating body, an imposed angle of heel causes the center of buoyancy to move sideways from its initial position on the centerline to a new position

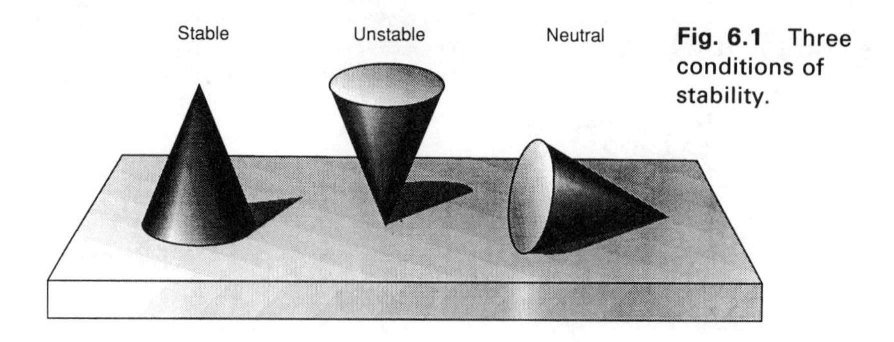

Fig. 6.1 Three conditions of stability.

somewhere in the more deeply immersed side. This is shown by the shift of B to B' in Fig. 6.2. The ship's weight (W) and buoyancy (or displacement) are of course equal, and both continue to act vertically even though the vessel is heeled. Please note that the values of both weight and buoyancy are symbolized by the Greek letter delta (Δ). In a stable ship the new center of buoyancy moves far enough to provide a turning moment (or couple) that will try to return the ship to the initial upright position. When that is true, the buoyant line of force will intersect the ship's centerline somewhere *above* the center of gravity. Naval architects call this intersection the *metacenter* because it is the

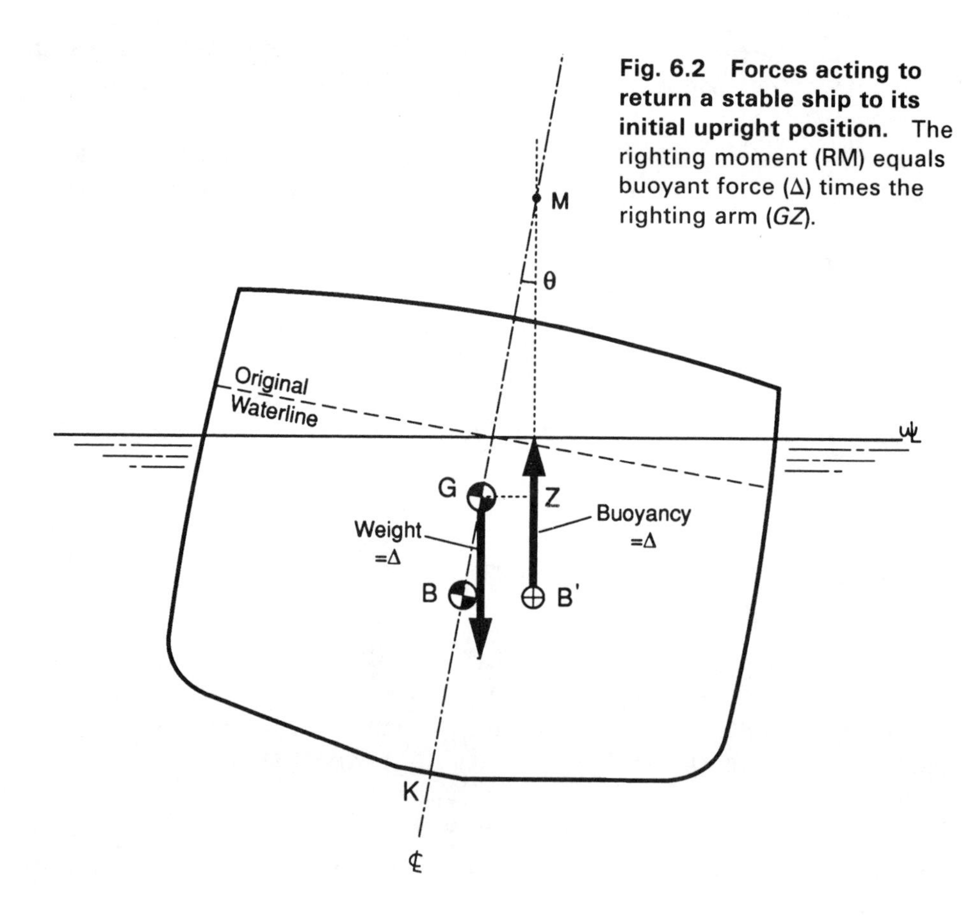

Fig. 6.2 Forces acting to return a stable ship to its initial upright position. The righting moment (RM) equals buoyant force (Δ) times the righting arm (GZ).

position at which the buoyancy *seems* to be acting. The metacenter's location is shown at Point M in the figure, and that is its standard abbreviation.

Buoyant stability is somewhat analogous to forces acting on a rocking chair (see Fig. 6.3). Again, the center of weight is higher than the center of support (which is down on the floor). When the person in the chair leans back, his or her weight moves, but the center of support also moves and so keeps the chair from tipping over. If you want to see how raising a weight can lead to instability, try standing up in a rocking chair—or a canoe (better wear a bathing suit for this latter experiment).

2.2 Metacentric height. Referring again to Fig. 6.2, suppose something were to raise the center of gravity from where it is shown (at *G*) to some location *above* the metacenter, *M*. Then, obviously, the turning moment would tend to *increase* the angle of heel and the vessel would be in a condition of negative stability. The distance between *G* and *M* is called the *metacentric height* and is considered positive if *G* is below *M*, or negative if *M* is below *G*. In plain words, positive stability is found whenever the position of the metacenter is above the vertical center of gravity, and negative stability arises when the positions are reversed.

The metacentric height is usually referred to simply as "*GM*." It is the key indicator of initial transverse stability.

The angle of heel is usually referred to by the Greek letter theta (θ). For ordinary hull forms, *GM* remains essentially constant for small values of θ, say up to 7 degrees. If greater angles are imposed, *GM* will at first tend to increase, but will then decrease and eventually become neutral, then negative, and a capsize will result This is discussed further in Section 2.9.

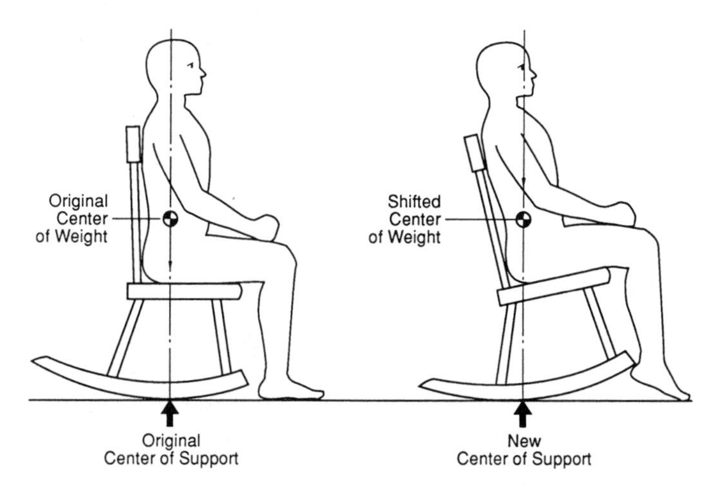

Fig. 6.3 The rocking chair analogy. As the occupant shifts his or her weight back, the center of support shifts to compensate.

2.3 Righting arm. The distance GZ shown in Fig. 6.2 is called the *righting arm* (or *righting lever*), and the *righting moment* equals Δ times GZ. At small angles of heel, GZ equals GM times the sine of the angle θ, that is, $GZ = GM \cdot \sin\theta$.

2.4 Predicting metacentric height. Clearly, the naval architect must know how to find GM. This is a three-step process. The first step is to find KB, the height of the center of buoyancy above the baseline (Fig. 6.2). An exact value can be derived from numerical analysis, the details of which are laid out in the references cited at the end of the chapter. For ordinary ship forms, KB will come out to be about 52 percent of the draft. Closer approximations recognize that fine lines (that is, low block coefficient) will tend to raise KB, as will also a V-bottom hull form.

The next step is to find the distance BM, which naval architects call the *metacentric radius*. It can be shown (see the references) that

$$BM = \frac{I}{V}$$

where I is the transverse moment of inertia of the ship's waterplane about its own centerline, and V is the volume of displacement.

Again, an exact value of I can be found by numerical analysis, as shown in the references. In the early stages of design, and before the lines are drawn, approximate values are found recognizing that I will be directly proportional to the ship's length multiplied by the cube of the beam. From this you can infer the importance of a wide beam in providing stability to your boat.

The naval architect's third task is to find KG, the vertical center of gravity of all weights: the empty ship plus all the deadweight items. Obviously, one never knows exactly how a ship will be loaded, so the figure for KG can never be more than an estimate. The prudent naval architect will assume it is rather high.

The naval architect now has all the ingredients needed to predict the metacentric height, GM. Referring to Fig. 6.2 you can see that

$$KM = KB + BM$$

and

$$GM = KM - KG$$

You may ask: Will stability be affected by moving from salt water to fresh? The lesser density of fresh water will allow the vessel to sink to a slightly greater draft, but the net overall effect on stability will be insignificant.

2.5 The inclining experiment. In an existing vessel, GM can be found by means of what naval architects call the *inclining experiment*. In this exercise a known weight (W) is moved from the vessel's centerline a distance d to one side (Fig. 6.4). The resulting angle of heel, θ, produces a lateral shift of the pendulum. How far will the vessel heel? It will heel until the righting moment (RM) exactly balances the heeling moment (HM). As shown in the references, this leads to the simple expression for the metacentric height:

$$GM = \frac{W \cdot d}{\sin\theta \cdot \Delta}$$

In practice naval architects substitute tan θ for sin θ. They do so because tan θ is more easily measured (as the distance the pendulum swings divided by the length of the pendulum) and sine and tangent are virtually equal at small angles. This leads to:

$$GM = \frac{W \cdot d}{\tan\theta \cdot \Delta}$$

You can try this experiment on your own boat. But be careful. Do it on a calm day. Make sure the mooring lines are slack. Get rid of all bilge water and other loose liquids, and make everyone on board hold perfectly still.

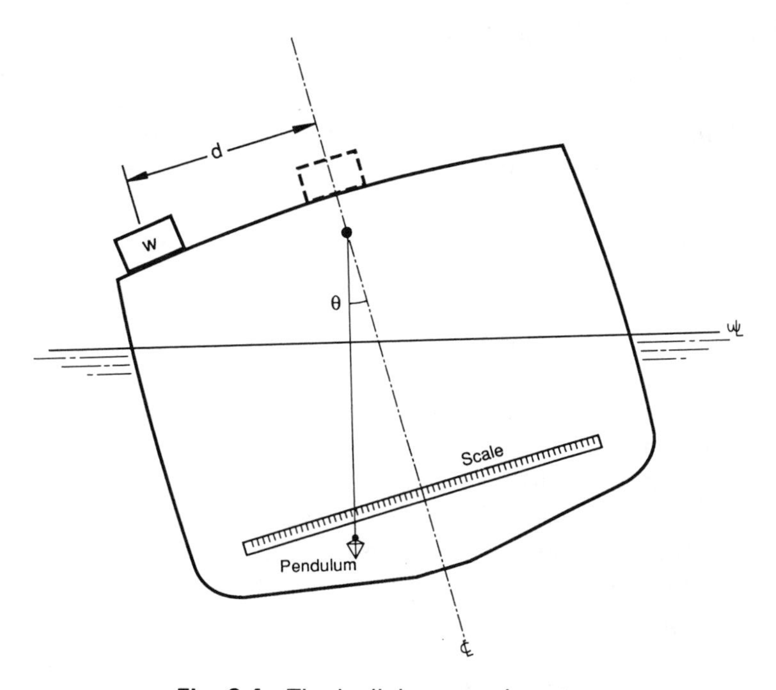

Fig. 6.4 The inclining experiment.

2.6 Ideal degree of stability. In designing a ship, the naval architect aims for a reasonable value of *GM*. Values of perhaps three to five percent of the beam are usually considered about right (much less than shown in Fig. 6.2). Smaller values leave too little margin in case of accident, careless loading, topside icing, or whatever. Greater values on the other hand, lead to excessive stability. Remember that when we say a ship is stable we usually mean it is *stable with respect to the water's surface*. Thus, a ship with great stability will automatically follow every wave profile that comes along. That will lead to short, uncomfortable rolling that may be dangerous to life, limb, and digestive system alike. Harsh rolling may also produce excessive stresses on rigging, smoke stacks, and other elevated objects.

Various regulatory bodies, such as the U.S. Coast Guard, issue rules pertaining to minimum acceptable metacentric heights. These apply to different kinds of vessels operating in different kinds of waters (sheltered, open, etc.). The naval architect interprets those rules and converts the results into readily usable charts or tables for the benefit of the vessel operator. These give the operator the upper allowable limits on the vessel's center of gravity (*KG*), depending on the draft.

2.7 Free-surface effects. An unpleasant surprise that can ruin stability may result from internal liquids of any sort. Assume such liquids are not snugly contained. If the ship heels to one side the liquids will flow in the same direction, thus shifting the ship's overall center of gravity toward the low side. This has the effect of reducing the metacentric height. Of course, if the internal liquid completely fills its tank, there will be no free-surface effect on stability.

Loose bulk cargos such as grain can easily shift, so their free surfaces must somehow be controlled if stability is to be maintained. In tankers, longitudinal bulkheads (usually two of them) are used to reduce the width of the cargo's free surface (they also reduce the danger of excessive dynamic forces resulting from liquids sloshing about).

2.8 Lifting a weight. Another unpleasant surprise may result from lifting a heavy object off the deck (Fig. 6.5). If the object is suspended from the end of a boom, as shown, the instant it is off the deck it is free to swing. When that happens its center of gravity jumps, in effect, from its actual center to an apparent center (that is, a metacenter) at the top of the boom. The shift in the ship's overall center of gravity, *KG*, will equal $(W \cdot d)/\Delta$.

2.9 Range of stability. To this point I have confined my exposition to stability characteristics at small angles of heel. What happens at larger angles? Usually the righting moment, $\Delta \cdot GZ$, will start to increase. It will reach a maximum at about the angle of heel at which the deck edge starts to immerse. Beyond that it will tend to decrease and eventually drop to zero. This critical angle is called the *range of stability*. At any greater angle the vessel will capsize and have a new

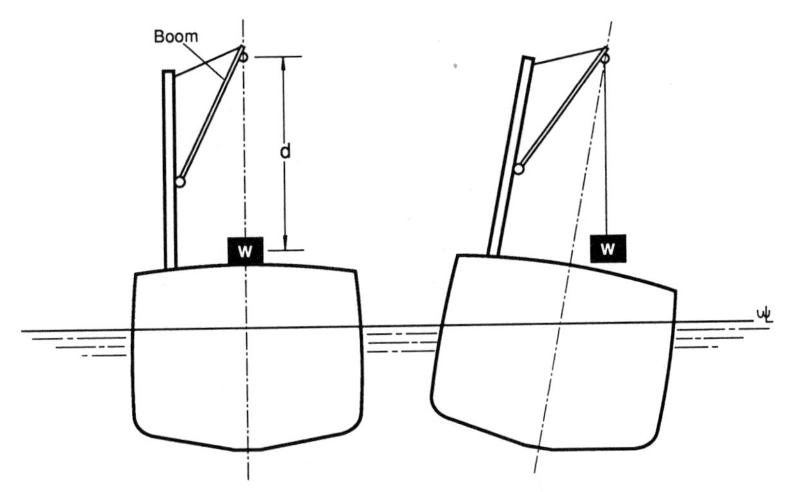

Fig. 6.5 Effect of shifting support of a weight (*w*) from deck to top of boom. As far as ship's stability is concerned, the weight might as well be hoisted to the top of the boom.

set of stability characteristics tending to keep it bottom side up. Figure 6.6 shows a typical curve of righting moments plotted against the angle of heel. Because the displacement remains the same at all angles, simply plotting *GZ* will be equally enlightening.

Righting moment curves such as those shown in Fig. 6.6 are somewhat unrealistic in that naval architects simplify the work by assuming fixed weights throughout. That is, they pretend that none of the cargo, equipment, stores, or people will slide downhill when the ship heels. In real life such sliding will exaggerate the heeling moment.

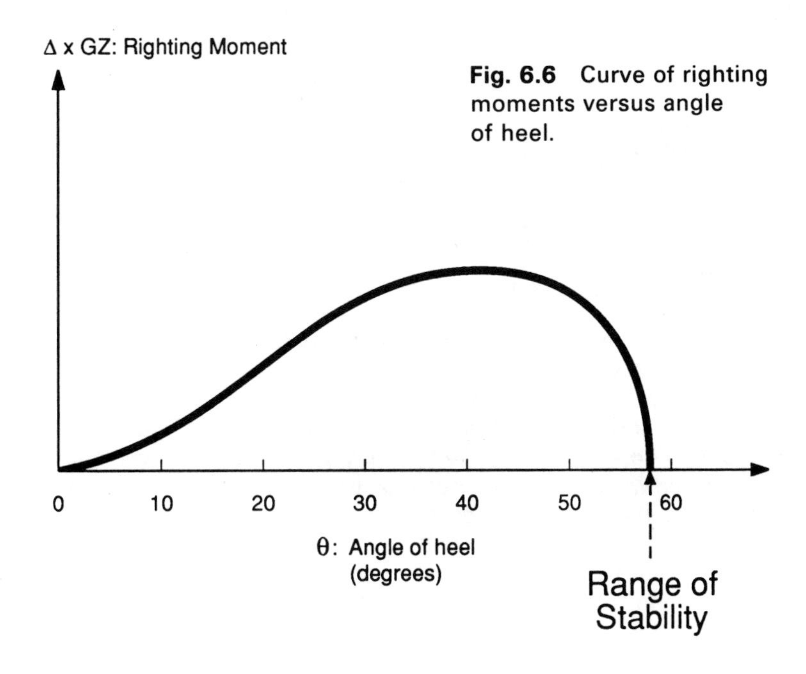

Fig. 6.6 Curve of righting moments versus angle of heel.

3. *Longitudinal Static Stability*

You need seldom worry about a ship capsizing in the fore-and-aft direction. You may, however, need to concern yourself with questions of trim as weights are added or removed from the ship, or shifted longitudinally. Specifically, you may find it useful to determine the vessel's susceptibility to trimming. The measure for this is MTI: the moment to trim one inch, or MCT 1 cm: the moment to change trim one centimeter. As you may recall, one or the other of these will be shown on the curves of form.

Figure 6.7 shows the forces at play as a weight is shifted from one end of a ship to the other (note the similarity to Fig. 6.2 in the section on *transverse* stability). As in the inclining experiment, the trimming moment, $W \cdot d$, will be balanced by the righting moment $\Delta \cdot GZ$. As explained in the references, the moment to change the trim one inch (or one centimeter) is directly proportional to the displacement multiplied by the longitudinal metacentric height and inversely proportional to the vessel's length. The longitudinal metacentric radius can be found through numerical analysis, and that, too, will be needed in the calculation. In practice naval architects are likely to use computer programs that do all this for them. All they need do is feed in the ship's length and the offsets of the waterplane at each station. The

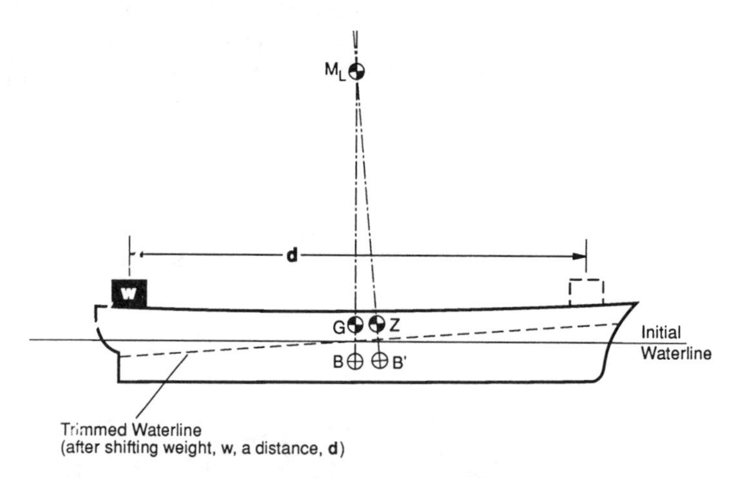

Fig. 6.7 Elements of longitudinal stability. Imagine a ship floating in still water with a weight at the stern. The center of buoyancy (shown as B) and center of gravity (shown as *G*) will be vertically in line. If the weight (*w*) is now moved a distance (*d*) toward the bow, the center of gravity will move to *Z* and the ship will trim by the bow until the new center of buoyancy (B′) is under *Z*. The lines of force through centers of buoyancy before and after the shift will intersect at M_L, the longitudinal metacenter. The trimming moment ($w \times d$) will be balanced by the righting moment ($\Delta \times GZ$).

computer, if properly programmed, will then do all the calculations and print out the results.

4. Flooding and Subdivision

4.1 Role of bulkheads. To enhance the safety of ships, naval architects try—as far as is practical—to incorporate numerous internal watertight divisions. These usually take the form of transverse diaphragms or walls, which are called *bulkheads*. This is done so that if the underwater hull is ruptured, the ensuing entry of water will be confined and not cause the ship to sink. (*Founder* is the technical term.) In naval vessels, decks as well as bulkheads may be made watertight, but in most other kinds of vessels only the uppermost, or weather, deck is so designed. Of course, most ships of any size (except perhaps tankers) have doublebottoms incorporating watertight innerbottoms that should confine flooding in case of bottom damage.

As of this writing there are few laws about bulkhead spacing except for passenger ships. Prudent shipowners, however, generally specify a "one-compartment" standard. This means that any one watertight compartment may be open to the sea without immersing the uppermost deck or allowing the ship to capsize. In one-compartment ships, one must hope that any future shell damage will not cross one of the watertight bulkheads. If it should do so, two compartments would flood and all bets would be off. To protect the ship against that eventuality, designers can provide two- or even three-compartment subdivision. As dictated by international rules, larger passenger ships must incorporate such higher degrees of subdivision.

4.2 Effect of flooding. What happens if the underwater shell is punctured and admits water into a watertight compartment? In effect, the buoyancy provided by that compartment will be lost. The ship will sink deeper into the water. Its longitudinal center of buoyancy will usually move and that will cause trim. The limiting length of a compartment (assuming one-compartment subdivision) is the maximum value that will confine sinkage and trim to such an extent that the uppermost deck will remain out of the water. Actually, to provide a modest allowance for error, naval architects aim to keep the deck at least three inches out of the water. Fig. 6.8 illustrates these points.

You may ask, "Why worry about immersing the deck? After all, the deck is watertight. Are spaces other than the damaged one likely to flood?" The answer is maybe yes, maybe no. The vulnerable feature is usually one or more hatch covers. (Hatches are openings in decks.) Hatch covers are *weathertight*, but not necessarily *watertight*. By this I mean that they are capable of shedding rain water or spray, but cannot always withstand any prolonged water pressure. It is true that

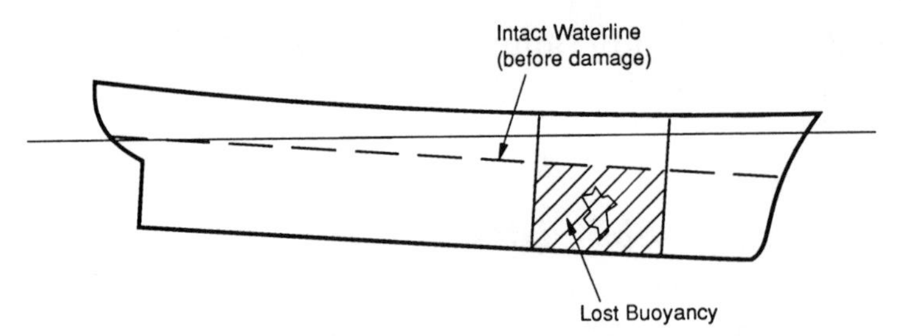

Fig. 6.8 Effect of flooding one compartment. The partial loss in buoyancy will allow ship to sink (and usually trim) until the net gain in displacement exactly offsets the loss. In a well-designed ship the deck will remain above water and the ship will not capsize.

most hatch covers are mounted on *coamings*, that is, low walls, that raise them perhaps three feet or a meter above the deck. It is also true that even if the hatch covers are clear of water, crew members cannot move about on the deck if it is under much water. This is especially true if waves are rolling aboard. As Kipling put it in *Poseidon's Law:*

> Save ye owe the Fates a jest,
> Be slow to jest with them.

Remember, too, that once the deck goes into the water, transverse stability starts to fall off rather rapidly. As explained in Section 4.4, a damaged, partially flooded ship is in danger of capsizing.

You may have noticed that tankers are allowed to operate with less freeboard than ordinary cargo ships. This is reasonable because tankers are finely subdivided by many bulkheads, both transverse and longitudinal, and they have small hatches (little more than manholes) with strong steel covers.

4.3 Subdivision standards. One of the more complex calculations in naval architecture leads to the ability to locate transverse bulkheads so as to provide one-, two-, or three-compartment subdivision. Needless to say, computers do most of this work today.

A ship is most vulnerable to foundering when it is deeply loaded down to its minimum allowable freeboard. If it be a cargo ship, this means the holds contain a quantity of cargo. That being the case, the volume of water that can enter a damaged compartment will be appreciably less than the compartment's molded volume. The loss in buoyancy will thereby be reduced. Naval architects must therefore make some realistic guesses as to the empty volume available for the entry of water. The ratio of that volume to the molded volume is called *permeability*. The standard assumed values are 85 percent for machinery spaces and 63 percent for ordinary cargo holds, 70 percent for container holds, and 85 percent for average RO/RO holds (with certain adjustments for passenger ships). These figures are obviously subject to error, but are usually on the safe side.

4.4 Stability after damage. Most of the foregoing discussion has centered on the desirability of keeping the uppermost deck out of water despite sinkage and trim. Of equal importance is the danger of capsize. Remember that the initial metacentric radius, *BM*, is proportional to the transverse moment of inertia of the intact waterplane. If a compartment is open to the sea its share of the intact waterplane is lost; *BM* will decrease and, as a result, so will the metacentric height, *GM*. A capsize may ensue.

5. Stability of Submarines

To this point everything I have said about stability has been confined to vessels floating on the surface of the water. When a floating vessel is heeled, a wedge of buoyancy shifts from high side to low. That is what causes the overall center of buoyancy to move toward the low side and thereby provide a righting moment. In heeling a completely submerged object, however, there are no shifting wedges. The center of buoyancy remains fixed; *BM*, in short, is zero. The same is true with respect to longitudinal *BM*. From this it follows that stability (whether transverse or longitudinal) demands that the center of gravity lie *below* the center of buoyancy. It also means that submarines are much more easily trimmed than surface ships. The weight of a person moving from one end of a submarine to the other will require compensation in some form to prevent an undesirable degree of trim.

There are some other unique problems in submarine design. See Goldberg (1988) for a complete discourse on submarine stability.

References and Further Reading

Barnaby, Kenneth C., *Basic Naval Architecture*, Hutchinson's, London, 1948.

Baxter, Brian, *Naval Architecture*, Teach Yourself Books, Hodder and Stoughton, London, 1959.

Comstock, John, *Introduction to Naval Architecture*, Newport News Shipbuilding Company, Newport News, Va., 1944.

Goldberg, Lawrence L., "Intact Stability," Chapter 2 in *Principles of Naval Architecture*, Edward V. Lewis, Ed., Society of Naval Architects and Marine Engineers, Jersey City, N.J., 1988.

Hamlin, Norman, "Ship Geometry," Chapter 1 in *Principles of Naval Architecture*, Edward V. Lewis, Ed., Society of Naval Architects and Marine Engineers, Jersey City, N.J., 1988.

Nickum, George C., "Subdivision and Damage Stability," Chapter 3 in *Principles of Naval Architecture*, Edward V. Lewis, Ed., Society of Naval Architects and Marine Engineers, Jersey City, N.J., 1988.

Rawson, K. J., and Tupper, E. C., *Basic Ship Theory*, Longman, London and New York, 1976.

Robb, Andrew, *Theory of Naval Architecture*, Charles Griffin & Company, London, 1952.

Fig. 6.9 Cargo ship in distress. This shows a U.S. Coast Guard helicopter rescuing seamen from a Soviet ship in danger of capsize and foundering off Cape Cod, March 1987. Courtesy R.W. Paugh, U.S. Coast Guard.

CHAPTER VII

DYNAMIC STABILITY

1. General Considerations

1.1 Perspective. In this chapter you will learn about the six types of motion in which a ship may find itself, particularly as it moves through waves. These motions are shown in Fig. 7.1. Three of the motions are rotational: rolling, pitching, and yawing. The others are translational: heaving (up and down), surging (fore and aft), and swaying (side to side). A ship may experience all six kinds of motions at the same time. Under certain conditions their interactive effects may be significant. Fortunately, however, that is rarely the case, which means that in most instances each kind of motion may be analyzed independently of the others.

The rotational motions are the most pronounced and troublesome, so those are the ones I shall try to explain in some detail. In each case I shall outline the causes of the motion and then discuss some of the practical methods used to minimize them.

1.2 Common features. Rolling and pitching motions are pendulum-like in nature. That is, given an initial displacement from the at-rest position, once the perturbing force is removed the ship will swing back to the normal at-rest position and keep on swinging in the opposite direction. This will continue until the restoring forces overcome the inertia forces that carried it past the normal position. Like a pendulum, were it not for friction and other inhibiting forces, the oscillations would go on forever. Yawing shares some of those same characteristics.

The most serious situations may arise if any sort of repeating perturbing force comes along at a frequency that coincides with the ship's natural period of rotation. This condition (called *resonance*) leads at least for a time to ever-increasing amplitudes and, under extreme conditions, may lead to a catastrophe. The most likely such situation involves a ship moving through beam seas (that is, with waves advancing at right angles to the vessel's course) with a frequency of encounter equal to the ship's natural period of roll. Under those conditions, even though the waves may be small, rolling may build up until the ship capsizes. Fortunately, nature seldom provides a single family of waves of exactly uniform periods. Moreover, there are energy-absorbing forces that tend to limit the degree of roll. Typically, the surface of the

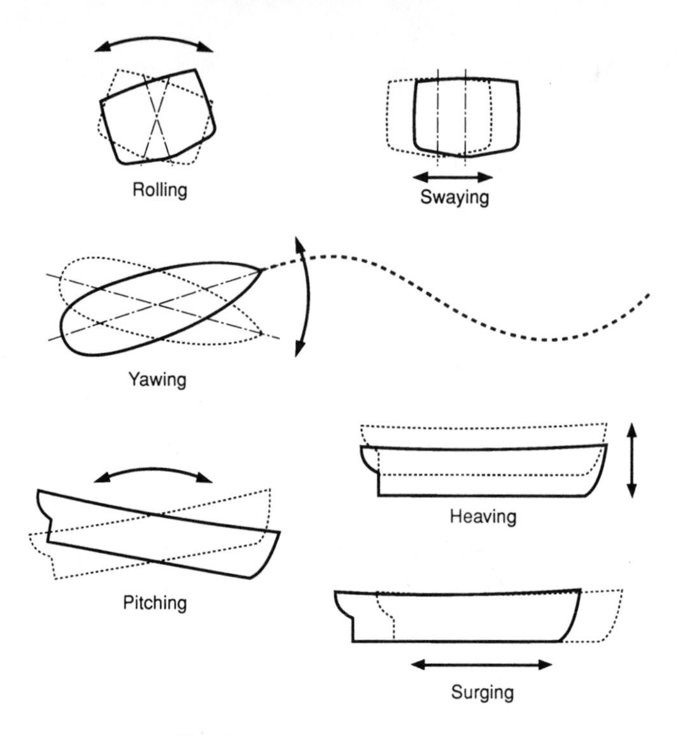

Fig. 7.1 Six types of motion.

sea is disturbed by several families of uniform waves, with each family moving at its own speed and in its own direction. At times two individual waves may come together to produce, momentarily, a single large wave. At other times one wave will momentarily cancel another. The overall result is the confused sea that one normally expects to encounter.

1.3 Sea-kindliness. The term *sea-kindly* is rather hard to define in quantitative terms. It implies physical attributes in a vessel leading to relatively benign motions (that is, modest in both magnitude and acceleration) and freedom from waves and spray coming aboard. Submerged submarines and vessels operating in calm, protected waters need experience no problems with motions. All others, however, will find some amount of motion inevitable. In the following sections I mention ways in which naval architects try to minimize those motions.

One must always consider the unhappy effects of motions on any human beings aboard. Working and living conditions both deteriorate. If the ship is tossing about, seafarers are less able to carry out their work and few passengers will be the least bit appreciative. Then, too, there is always the threat of falls leading to sprains or broken bones. Naval architects seek guidance in matters pertaining to human tolerance for ship motions. Mandel (1989) is such a source.

Once the vessel goes into service, judicious operation can help a good deal. The operators, for example, can minimize the ill-effects of waves by changing speed or course, or both. Modern weather-routing services can also help by allowing the officers to modify routing so as to escape unfavorable sea conditions. I mention other ways operators can help reduce motions as we go into more detail below.

2. *Rolling*

2.1 Causes. I have already mentioned the principal cause of rolling: beam seas. Other causes include variable winds (of particular concern in sailing craft), and such occasional forces as those applied by shifting cargo or putting the rudder hard-over while moving ahead at speed. In a small craft, a person may start the boat rolling simply by moving from one side to the other.

2.2 Disagreeable aspects. In addition to the possibility of capsizing the ship, rolling carries other threats. Any person or object far from the center of motion would experience induced loads. Inertial forces reach a maximum as the vessel reaches the end of a roll, pauses, and starts to reverse. Masts, smokestacks, and other elevated parts of the vessel tend to continue their transverse motion at that point. If not properly secured they may go right on over the side.

I must also point out that problems with seasickness are more often associated with rolling than with any other form of ship motion.

> There's a nasty up-and-down motion,
> That comes from the bed of the ocean,
> And kinda gives me the notion
> I was never meant for the sea!
> (Author unknown)

2.3 Cures. As we mentioned in the preceding chapter, excessive initial static stability leads to harsh rolling because the vessel tends to follow every wave profile that comes along. Modest values of the metacentric height, *GM*, are therefore appropriate. This is more easily provided for in large ships. In small craft modest values of *GM* may not allow sufficient safety margin because a relatively modest shift of weight may cause a capsize.

Most large ships are fitted with what are called *bilge keels* (Fig. 7.2). When a vessel rolls, the keels move large masses of water and also create turbulence. This acts to resist rolling and so reduces the amplitude. The same tendency will show up in ships with square bilges.

Some ships are fitted with antiroll tanks as shown in Fig. 7.3. When properly designed, this system produces an oscillating transverse flow of water so timed as to generate loads that are opposite to the perturbing force. Simply stated, they add weight to the high side of the

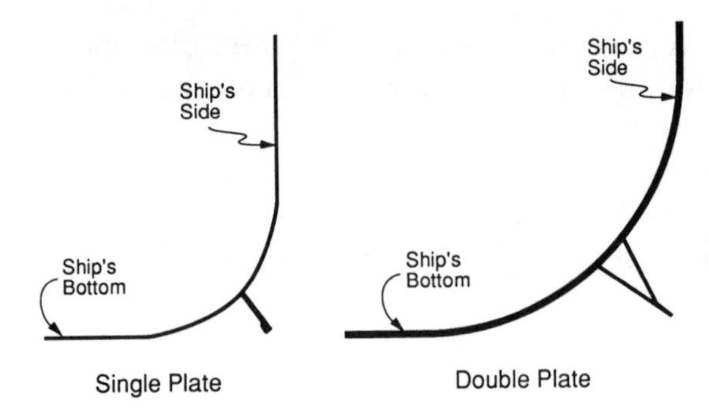

Fig. 7.2 Bilge keels. The double-plate type can be made much wider than the single-plate type. It is suited to ships with generous bilge radii.

ship when it rolls. This will not stop rolling, but can markedly reduce its amplitude. Their disadvantage lies in the not-inconsiderable weight of water, and corresponding loss in useful deadweight.

Another approach is to fit antiroll fins port and starboard at the turn of the bilge as shown in Fig. 7.4. These fins are on pivots and are controlled by sensors that can tell which way the ship is rolling. These sensors cause the fins to pivot in such a way as to counteract the roll. The fins are turned so that, as they move ahead through the water, the

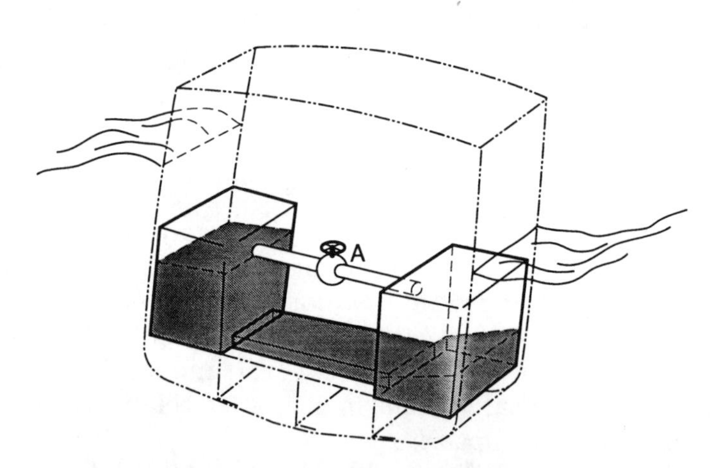

Fig. 7.3 Antiroll tank. In this arrangement the contained water, in surging from one side of the ship to the other, reduces the amplitude of roll. It does this because, once set in motion, the water's momentum tends to carry it to the vessel's high side. An air valve is sometimes fitted at Point A. This can be adjusted to change the natural surge period of the internal water to suit the ship's loading condition. Notice that the weight of water in the tanks opposes the direction of roll.

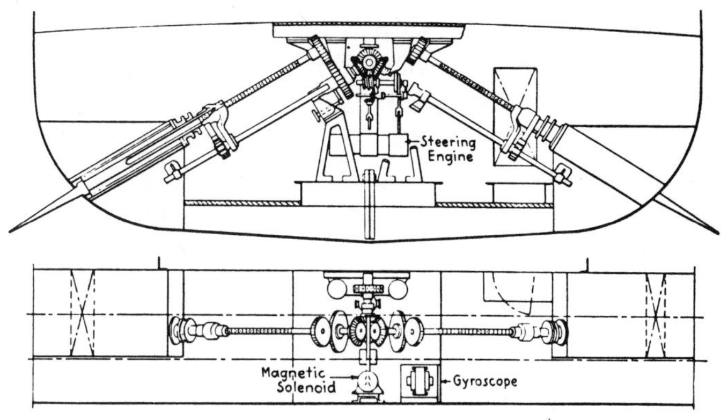

Fig. 7.4 Fin stabilizers (source: Lewis, 1967).

one on the low side provides an upward thrust and the one on the high side provides a downward thrust. To prevent damage while docking, most fins can be retracted into the hull. Obviously, this system is at its best only when the ship is moving ahead; it is of little use when at anchor.

As you probably have observed, a spinning gyroscope has a strong tendency to maintain the orientation of its axis. When fitted into a vessel, a spinning gyroscope will resist applied rotational forces. This is why some yachts have been fitted with large gyroscopes. Back in 1932 the Italian passenger liner *Conte de Savoia* was placed in service with three such massive devices. They were heavy, voluminous, expensive, and only marginally successful. The idea has never really caught on.

An earlier and bolder attempt was Henry Bessemer's swinging saloon steamer intended for the Dover-Calais ferry service. Mr. Bessemer was much prone to *mal-de-mer* and so aspired to offer to the public a ship with the major passenger accommodations carried in a giant steel box mounted on pivots fore and aft. The idea, of course, was to allow the ship to roll, but to keep the box level through manually operated hydraulic controls. The ship was placed in service in 1875, but the practical complications were such that the antirolling feature was never made operable. Failing as a passenger ferry, the ship was rebuilt and converted to cargo service, but ran aground almost immediately and was subsequently scrapped. With more sophisticated technology now available to us, perhaps the time is ripe to reinvent the concept.

I mentioned that rudder action can cause a roll. That effect can be turned around, and steering systems can be adapted so as to allow rudders to be used as roll-reducers.

A boat's period of roll is directly proportional to its *radius of gyration,* which is a measure of the boat's distribution of mass about its axis of rotation. The greater the radius, the greater the resistance to change in rotational motion. In practical terms this means that you can slow your boat's rolling period by moving weights symmetrically outboard, away from the center line.

If you find your boat is rolling excessively, a change of course may move it away from synchronous wave encounters. In larger vessels, in addition to changing course, partially filling a ballast water tank may help. That will probably lower the ship's overall center of gravity, but the metacentric height, *GM*, will be reduced because of the free surface effect. Rolling may thereby be diminished.

Ships designed to carry extremely dense bulk cargoes, such as iron ore, usually have relatively deep innerbottoms. This raises the ship's overall center of gravity, thus avoiding an excessive *GM* (metacentric height), thus reducing the severity of roll.

2.4 *GM* versus roll period. It can be shown in theory that a ship's natural period of roll can be closely predicted by this relationship:

$$T = \frac{1.108K}{\sqrt{GM}}$$

where

T = period for one complete cycle, that is, over and back, seconds
K = ship's transverse radius of gyration, ft
GM = ship's transverse metacentric height, ft

The value of K is closely related to the ship's beam, B. This leads to the useful approximation

$$T = \frac{0.42B}{\sqrt{GM}}$$

(The constant in the numerator may vary from 0.37 to 0.60, but 0.42 is a good average figure for normal designs.)

You may rearrange the expression. Then, having timed your boat's natural rolling period, you can approximate the metacentric height thus:

$$GM = \left(\frac{0.42B}{T}\right)^2$$

3. *Pitching*

3.1 Causes. Pitching is caused almost exclusively by the vessel running through waves. Less frequently, it may occur as an icebreaker's bow rides up on an ice cover and then drops as the ice is broken. This is one case where pitching is looked upon as a virtue.

3.2 Disagreeable aspects. Pitching often leads to taking seawater on deck, with threat of damage to personnel, the ship itself (particularly hatches and doors), or any cargo carried on the open deck. In extreme cases, entire deckhouses have been swept overboard.

The *forefoot* is the part of the ship where the keel rounds up into the stem. If, as often happens, it should emerge during periods of heavy pitching, then the bottom structure may be damaged as the ship slams into the water on its downward pitch. At the other end of the ship, the propeller may emerge from the water in whole or in part. With such exposure, the load torque will immediately fall off and the suddenly underloaded propulsion machinery may run overspeed and damage itself.

Once again inertia forces will be induced, just as in rolling. Now, however, because the radius of swing is so much greater, the resulting inertia forces become appreciably more severe. Cargo stowed near bow or stern may be crushed and cargo-carrying decks may collapse. Accelerations may exceed that of gravity, leading to cargo (or people) being tossed about and damaged.

3.3 Cures. In operation, usually the most effective cure for excessive pitching is to change course or speed, or both. Also, in loading fragile cargo at the ends of the ship, the upper layers should be braced against bouncing up, using various kinds of packing material such as rough lumber. In addition, where possible, the heavier cargo should be carried near amidships, thus reducing the longitudinal radius of gyration. That will allow the vessel's ends to ride up more easily when encountering large waves. For the same reason, a small boat will ride more safely in rough water if everyone sits low and near the middle.

Horizontal fins, whether in bow or stern, have been tried as a means of reducing pitch, but without great success. They tend to produce undesirable vibrations and are susceptible to damage if the extent of pitching is enough to allow them to emerge and then slam back into the water.

Planing craft intended for open-water operations should be given a pronounced deadrise, approaching 30 degrees, near the bow.

Naval architects try to keep the inevitability of pitching in mind when designing a ship. To help keep water off the deck, they may provide plenty of freeboard at the ends of the vessel, particularly at the bow. An outward flare of the shell near the bow will help throw water away from the ship, and structural deflectors (*breakwaters*) fitted near the bow on the weather deck can help prevent unrestrained tons of water from sweeping aft. In small craft, narrow longitudinal appendages (called *spray strips*) on the forebody shell will help keep waves and spray from coming aboard. If the boat has chines at the bilges, these should be swept up at the bow so as to be above the top of the bow wave when under way. Finally, if the after hull form is kept much like the forward end (that is, avoiding broad, flat underwater stern sections), then following seas will be less likely to cause heavy pitching.

Naval architects try to avoid unusually high values of the length-to-draft ratio. That will minimize the frequency of forefoot and propeller emersion. If such emersions cannot be avoided, forefoot damage can be minimized by incorporating V-shaped (rather than U-shaped) transverse sections in the bow. Engine over-speeding may be prevented by detection devices that will throttle back the engine whenever the propeller emerges.

Naval architects design extra strength into cargo-carrying decks and tank tops at the ends of the ship so as to cope with the large inertia forces likely to be experienced at those locations.

4. Yawing

4.1 Where it occurs. Yawing is most often observed in towed vessels, such as barges or, in the yachting world, in tenders (rowboats towed astern). You can readily understand why yawing sometimes becomes a problem. A vessel towed at the end of a long line may come along without trouble until the first perturbation causes a slight swing to either side. When that happens the water impinging on one side of the bow may force an ever-greater swing until the pull of the towline brings the forces into momentary balance. Now the towed vessel will swing back and head for the original course. But, having returned to course it will now be headed off in the opposite direction. After a few such cycles a sort of stable yawing condition will prevail and the swings will get no more extreme.

4.2 Disagreeable aspects. If a towed barge, or other vessel, is allowed to yaw well off course, its towline resistance will not lie along the tug's centerline. Its pull will be directed off at an angle, one component of which will tend to capsize the tug. The same effect is easily seen in waterskiing: speedboats are sometimes overturned by overzealous skiers. Clearly, too, there is danger that a yawing barge may run aground, collide with other vessels, or damage buoys and other aids to navigation.

A vessel running in the direction of the wind and waves will experience following seas (waves tending to overtake the vessel from astern). These may cause the vessel to yaw 90 degrees off course, throwing it into beam seas. This condition is known as *broaching* and in extreme cases it may produce a capsize.

4.3 Cures. A trim by the stern will cure, or at least reduce, yawing regardless of vessel size. If your yacht's tender yaws badly, adding some weight in the stern may solve the problem. If a crewed ship with operable rudder is being towed, the helmsman can prevent yaw. Various techniques may be incorporated to prevent yaw in towed, unmanned barges. The most common method is to fit twin stabilizing fins, called *skegs*, near the stern (Fig. 7.5).

In the section on pitching, just above, I said that designers are well advised to avoid broad, flat underwater stern sections, because they may lead to excessive pitching. Another reason to avoid them is that they tend to promote broaching. A deep forefoot should also be avoided if broaching is likely to be a problem.

5. *Closure*

I have told you about the more important kinds of ship motions, their causes, and their detrimental effects. I have also outlined ways, both in design and in operation, to help minimize motions. Although there are many relevant, preventive steps that can be taken, total elimination of detrimental motion seems all but impossible except when a vessel is operating in placid waters or is fully submerged. Perhaps the best we can hope for is a ship or boat that will be gentle in its motions and little prone to allowing green water or clouds of spray on deck. As already mentioned, naval architects have a term to describe such an ideal craft. They say it is sea-kindly.

To close this chapter on a practical note, let me say a few words about preventing seasickness. The central villain in causing any form of motion sickness is supposedly the impact of the motion on one's diaphragm. I don't understand them, and so I shall not attempt to explain the physiological effects or psychological influences that bear on the problem. I can, however, call to your attention some common-sense actions you can take to minimize the forces on your diaphragm. First, lie down. Second, and particularly if you are not satisfied to spend your cruise supine, get yourself to that part of the ship where the up-and-down motions are minimum. That would be at that point

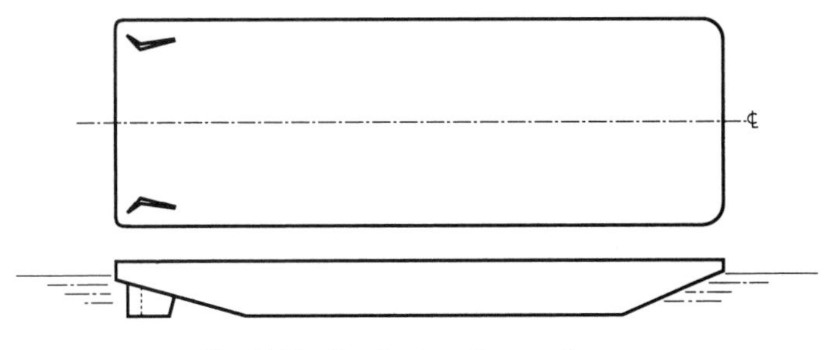

Fig. 7.5 Typical antiyaw skegs.

central to both rolling and pitching. The rolling center is usually close to the waterline and, of course, on the vessel's centerline. The pitching center is likely to be somewhat aft of mid-length and, again, near the waterline. In my own experience, being in fresh air is a help, but I am in no position to explain why. Psychology is often an important element. Try to keep your mind off the problem and discourage others from talking about it.

Naval architects can help if they keep unavoidable motions in mind when blocking out the vessel's arrangements. That is, they should try to place the living accommodations and key working spaces (principally the navigating bridge) in locations where motions should be a minimum.

If all else fails, see a doctor about medicines or skin patches that may help; or you might want to try iceboating.

References and Further Reading

Beck, Robert F., Cummins, William E., Dalzell, John F., Mandel, Philip, and Webster, William C., "Motions in Waves," Chapter XIII in *Principles of Naval Architecture,* by Edward V. Lewis, Ed., Society of Naval Architects and Marine Engineers, Jersey City, N.J., 1989.

Crane, C. Lincoln, Eda, Haruzo, and Landsburg, Alexander C., "Controllability," Chapter IX in *Principles of Naval Architecture,* Edward V. Lewis, Ed., Society of Naval Architects and Marine Engineers, Jersey City, N.J., 1989.

Hamlin, Cyrus, *Preliminary Design of Boats and Ships,* Cornell Maritime Press, Centreville, Md., 1989.

Mandel, Philip, "Assessing Ship Seaway Performance," Section 7 of Chapter VIII in *Principles of Naval Architecture,* Edward V. Lewis, Ed., Society of Naval Architects and Marine Engineers, Jersey City, N.J., 1989.

Källström, Claes G., Wessel, Peter, and Sjölander, Peter, "Roll Reduction by Rudder Control" in *Proceedings,* 13th Ship Technology and Research (STAR) Symposium and 3rd International Marine Systems Design Conference (IMSDC), Society of Naval Architects and Marine Engineers, Jersey City, N.J., 1988, pp. 67–76.

Lewis, E. V., "The Motion of Ships in Waves," Chapter IX in *Principles of Naval Architecture,* John P. Comstock, Ed., Society of Naval Architects Marine Engineers, Jersey City, N.J., 1967.

Fig. 7.6 Winter North Atlantic. Two views of the U.S. Coast Guard cutter *Mendota* on ocean station east of Nova Scotia, February 1965. Severe pitching such as this imposes heavy loads on the ship's bottom structure. Courtesy R.W. Paugh, U.S. Coast Guard.

Fig. 7.7 Antiyaw skegs. The two appendages hung under the stern of this barge are effective in keeping the towed vessel on course. The flaps at the trailing edges can be turned back when the barge is being pushed, rather than pulled, thus reducing resistance when not needed to control yaw. Courtesy Dravo Corp.

CHAPTER VIII

RESISTANCE AND POWERING

1. *Historical Perspective*

Until somewhat over a century ago, predicting the speed of a proposed ship was all art and no science. Then, as steam engines replaced sails, half the difficulty was removed: naval architects no longer had to guess what wind forces and directions would be available to push the ship along. The other half of the difficulty lay in predicting the resistance of any given hull form to being moved through the water at any given speed. Naval architects still have no precise quantitative understanding of that problem. Since the early 1870s, however, the insights of William Froude have produced generally satisfactory ways to estimate resistance. These are usually based, either directly or indirectly, on the use of model basins, which are in effect large analog computers. You will learn more about model basins in Section 4.

In summary of the above, naval architects still have a lot to learn about predicting the speed and power of boats and ships. Their estimating methods are based on perfectly sound principles. The numbers used, however, are nearly all derived from empirical evidence, either from model tests or from existing ships. Adding to the difficulty are hard-to-predict influences such as sea conditions, loading conditions, and degree of fouling (for example, barnacles) on the underwater hull.

In the rest of this chapter I shall outline rational ways to analyze the speed and power relationship for normal displacement type hull forms. We shall leave the problems of planing hulls, hydrofoil craft, and air-cushion vehicles to more advanced references.

2. *Components of Resistance*

2.1 Submarines. Let me start this discussion of the components of resistance by considering the case of the deeply submerged submarine. Let us assume that the craft is well shaped. That being the case, the only important source of resistance to forward motion will be the friction caused by the viscosity of the water. The energy used to force the water aside at the bow will be fully regained as it closes together at the stern.

2.2 Friction. Contrary to popular opinion, a ship's frictional resistance is not so much between hull and water as it is within the water itself. Let me explain. As the hull slides through the water, a thin coat of the water will attach itself to the the hull and move along at the ship's speed. Owing to the water's viscosity (stickiness), this innermost layer will try to induce the second layer to come along; but that same viscous property will cause the third layer to resist the forward motion of the second layer. The cumulative result is that many layers will be caught up and tend to follow the hull, but each will be less affected than its interior neighbor. Eventually, as the effect gets smaller and smaller, it disappears altogether at some finite distance from the hull. The body of water that is dragged along is known as the *boundary layer.* Its thickness will be influenced by the relative roughness of the hull surface and will gradually increase from bow to stern. The accumulation of trailing water at the stern is the *wake.*

So, while we can reduce frictional resistance by keeping the hull clean and smooth, we only fool ourselves by coating it with grease, a trick that has been reinvented a thousand times but never with success.

2.3 Wave-making. If our submarine rises to the surface, a new source of resistance will appear: wave-making. The force of the bow pushing against the water will now produce waves. Since the waves must overcome gravity, they will absorb energy. Of course there are other waves created, too, such as those at the stern. We can generalize by saying that wave-making resistance arises from variable pressures between hull and water surface working against gravity.

Another source of resistance closely related to wave-making is that of eddies forming behind blunt appendages, shaft struts, rudders, and so forth. In a well-designed vessel this source of resistance is not large and is usually lumped in with wave-making. For reasons to be explained in Section 4, the wave- and eddy-making resistances are often referred to collectively as *residual resistance.*

2.4 Major components. Most ships and boats, then, have two major components of resistance to forward motion: friction (overcoming viscosity) and residual (largely wave-making, overcoming gravity). Naval architects try to minimize frictional resistance by providing a smooth hull and then trusting to the operators to maintain it in good condition. This requires frequent cleaning and the application of high-quality paints to resist both fouling and corrosion. Naval architects try to minimize residual resistance by developing good hull forms.

In most merchant ships, the frictional resistance is perhaps three to four times as large as the residual resistance. In relatively high-speed vessels (such as passenger ships, naval combat craft, yachts, or ferryboats), the two resistances might be about equal. Except in a small minority of cases, air resistance is so small (perhaps only a percent or two of total resistance) that naval architects feel safe in ignoring it altogether.

3. *Estimating Horsepower*

3.1 Central aim. Among a naval architect's most challenging responsibilities is that of selecting an engine that will drive a proposed ship at the right speed. Once the vessel is placed in service, failure to meet the owner's specified speed will cause the designer acute embarrassment. Worse yet, it may lead to extensive employment for members of the bar. On the other hand, if the vessel is found capable of appreciably greater-than-specified speeds, the embarrassment arises from having wasted the owner's money on an oversize engine. Hitting the bull's eye is no easy task.

3.2 Procedure. In selecting an engine the naval architect must go through a multistep procedure culminating in an estimate of the required horsepower: brake horsepower (BHP) if a diesel or gasoline engine, or shaft horsepower (SHP) if a steam turbine. BHP and SHP are identical in a direct-connected diesel engine. In geared engines BHP goes into the reduction gears; SHP comes out. The difference, caused by friction within the gears, is usually less than 3 percent.

Explaining the procedure for estimating power is easier if we start with the engine output (BHP or SHP) and work our way backward to see what happens to that power and how it is finally dissipated.

3.3 What happens to the energy. Some of the energy that goes into the propeller shaft never reaches the propeller. A small fraction (perhaps 1 or 2 percent) goes into overcoming friction in the shaft bearings and the seals that keep water from seeping in where the shaft passes through the hull. Propellers convert torsional energy into thrust. In doing this they are seldom more than 70 percent efficient, so the useful energy coming out of them is appreciably less than what went in. As you will learn in the next chapter, propeller efficiency can be readily estimated from published charts derived from numerous model propeller tests. What is more difficult, however, is to assess the complications brought on by interactions between the propeller and the hull.

I explained in Section 2.2 how friction produces a body of water, called the wake, that moves along with the ship. The propeller operates within this relatively slow moving water. That gives the propeller a firmer base for producing thrust and so it regains some of the energy lost in overcoming friction. At the same time, in drawing water into its sweep, the propeller reduces the water pressure on the stern. That "hull suction" (technically called *thrust deduction*) gives the propeller more work to do and so more energy is lost.

Naval architects generally lump the wake gain and thrust deduction to produce an overall estimate of their combined effect called *hull efficiency.* Typically, this hull efficiency may come to 1.10 in a single-screw ship and 1.00 in a twin-screw ship.

3.4 Effective horsepower. The product of propeller efficiency and hull efficiency is called the *propulsive coefficient.* This is used to convert the power delivered to the propeller to the power that would be required to pull the ship through the water if no propeller-related complications were involved. This is sometimes called *towrope horsepower,* but more commonly *effective horsepower* (EHP). It can be estimated from the three independent components of resistance already explained: friction, wave-making and a small increment for appendages. Let me discuss frictional resistance first.

3.5 Estimating frictional resistance. Numerous experiments have shown that frictional resistance will vary directly with the ship's wetted surface (which is the surface area of the underwater hull), with the ship's speed raised to an exponent somewhat less than two and, of course, the degree of roughness of the hull. Other factors are the ship's length and the viscosity of the water, which will vary slightly with temperature and degree of salinity. Long ships have relatively less frictional resistance per unit of area.

3.6 Estimating frictional horsepower. Once naval architects have estimated the frictional resistance (R_f), they can convert that figure into the frictional component of EHP, which is abbreviated EHP_f, by means of the relationship

$$\mathrm{EHP}_f = \frac{R_f(\text{in pounds})\cdot(\text{speed in feet per minute})}{33\,000}$$

or

$$\mathrm{EHP}_f = \frac{R_f(\text{in pounds})\cdot(\text{speed in knots})}{325.7}$$

Figure 8.1 shows frictional horsepower per thousand square feet of wetted surface versus speed, with contours for ships of different lengths. They assume salt water at 50°F and a degree of roughness typical of a new, carefully painted ship.

3.7. Estimating wave-making resistance. The task of predicting wave-making resistance is considerably more difficult. Naval architects simply do not understand in any quantitative way the physics of how a ship creates waves. Strictly mathematical approaches to the analysis have been under study for several decades, but most design methods in use today still rely on empirical evidence from model basin work (as explained in Section 4) or derive their conclusions from measured performances of existing similar ships. In some instances, model basin researchers have tested large numbers of methodically related hull forms. These are called *standard series,* and their results have been analyzed and published so that naval architects can derive

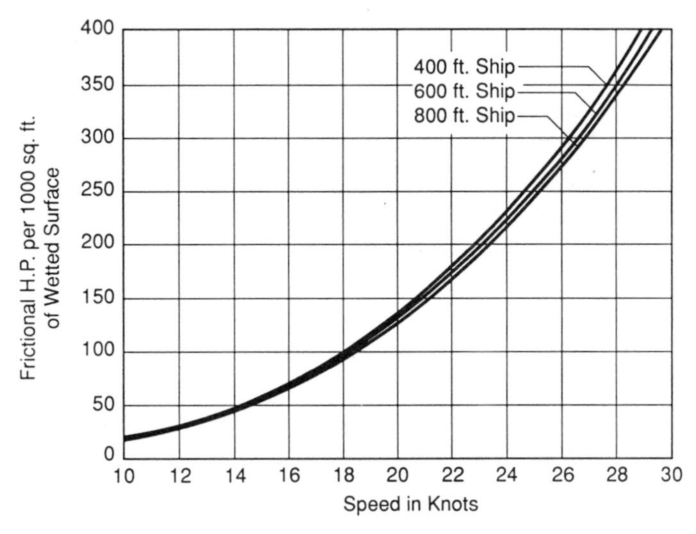

Fig. 8.1 Frictional horsepower per 1000 square feet of wetted surface, at various speeds, and for ships of various lengths (source: Comstock, 1944).

a good approximation to the wave-making resistance per ton of displacement for any normally shaped hull form. Figure 8.2 shows a typical set of residual (that is, wave-making) resistance contours from such a standard series. What is shown is for just one set of assumptions about the hull form and relative speed: namely, that the beam-draft ratio is equal to 2.25 and the speed-length ratio equals 1.00. Other charts would show contours appropriate to many other sets of assumptions.

Once wave-making resistance has been estimated, wave-making horsepower can be derived exactly as was shown for relating frictional resistance and frictional horsepower. The sum of the two horsepowers, plus a modest increment for appendages, will produce the total effective horsepower.

3.8 Recapitulation. To recapitulate, required engine powers are usually estimated by the following step-by-step procedure:

1. Measure the proposed vessel's wetted surface from the lines drawing and use that to estimate frictional horsepower.

2. Estimate wave-making resistance and corresponding horsepower requirements from published standard series results or from available data on existing similar ships. If time and budget permit, the services of a commercial model basin may be engaged not only to check estimates, but perhaps to refine the hull form by testing variations on the original design.

3. To the sum of the frictional and wave-making horsepowers add an increment for appendages that were not fitted to the model. This may amount to about 7 percent for a single-screw ship with a rudder

as the only appendage. In a well-designed twin-screw ship it may come to 12 percent.

4. The sum of the powers found in the first three steps will be the estimated effective horsepower, EHP. Dividing that figure by the propulsive coefficient leads to the horsepower that must be delivered to the propeller. The naval architect then adds a few more percentage points for shaft losses to arrive at SHP. Another small increment will provide BHP.

5. Having carefully developed estimates of SHP or BHP (often to four significant figures), the naval architect must now add a prudently generous margin in recognition of real-life conditions the ship will have to face. To this point all calculations have been based on an idealized set of assumptions: perfectly smooth hull, fair weather, and smooth seas. The usual margin for service conditions is a rather arbitrary 15 to 25 percent. Another margin may be appended if the type of engine is such that its output may be expected to diminish over the years. I think you may now see why this procedure for arriving at the necessary horsepower should be called an *estimate* and not a *calculation.*

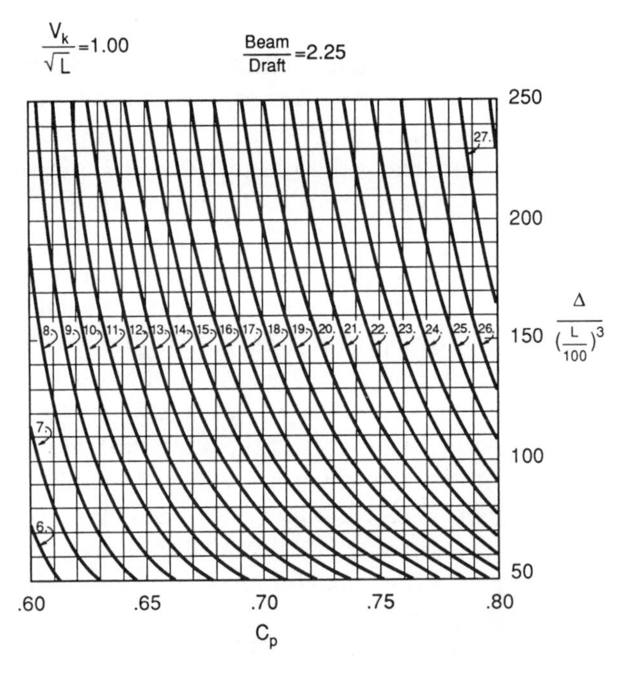

Fig. 8.2 Contours showing residuary resistance in pounds per ton of displacement for one family of standard hull forms with beam/draft ratio of 2.25 and speed/length ratio of one. Example: If the prismatic coefficient, C_p, is 0.70 and the displacement/length ratio is 200, the residuary resistance per ton of displacement would be 19 lb.

6. With the above-derived estimate of required power, the naval architect is ready to select a standard engine or turbine, usually right out of some manufacturer's catalog. Since off-the-shelf units will seldom provide the exact power required, prudence dictates going for a standard model of slightly more power than the estimate.

4. *Model Basin Theory*

4.1 Froude's insight. Before Froude's basic work, starting in 1868, various engineers (including Benjamin Franklin) had tried to predict ship speed based on model tests. They had all failed, however, because they did not understand how to extrapolate the model results to the full-size ship. Froude realized that the difficulty lay in the fact that the two major components of resistance (friction and wave-making) followed different physical laws in scaling up from model to ship. One of his unique contributions, then, was to treat friction and wave-making separately in going from model results to full-scale prediction.

Froude understood the laws of similitude. These tell us how physical characteristics are affected by scale. A simple example is the area of a circle. What happens to the area if we double the radius? Since, as I trust you will recall, Area $= \pi r^2$, we can see that doubling the linear dimension (that is, the radius) will increase the area by a factor of four. That is, we say that area varies with the scale squared. Similarly, volume varies with the scale cubed. If the ship is 20 times longer than the exact scale model, its volume of displacement will be that of the model multiplied by $20 \times 20 \times 20$, or 8000.

In Chapter V, as you may recall, I explained Froude's number:

$$FN = \frac{v}{\sqrt{g \cdot L}}$$

Froude observed that when ship and model were both moved at the same Froude number both produced identical (to scale) wave profiles. That is, if the model showed a wave crest at the bow and another at the stern, the ship's wave profile would show exactly the same pattern. From this he reasoned, correctly, that ship and model wave-making resistances would vary directly as their displacements when both were moved at the same Froude number.

As for frictional resistance, simple towing tests on a series of flat planks (essentially no wave-making) showed him that the viscous resistance varied directly with the wetted area and with the speed raised to a power somewhat less than two.

4.2 Froude's technique. Armed with the concepts spelled out above, Froude was ready to use his model basin. He had built an exact scale model of some ship. He then towed it through the water of his tank at the Froude number corresponding to the ship's design speed, and measured the pounds of pull required to maintain that model speed. This gave him the model's total resistance. He then calculated how much of that total was contributed by frictional resistance. Subtracting that from the total resistance gave him the (wave-making plus eddy) resistance. And now you know why the cumulative wave and eddy component is called *residual resistance.*

Now Froude was ready to use his model results to predict the full-scale ship's resistance (and from that, EHP). First, the ship's frictional resistance was predicted using much the same formulation as that used for the model, with a correction for relative roughness. Second, the ship's residual resistance was taken as the model's residual resistance multiplied by the ratio of their displacements (with, of course, ship and model speeds both at the same Froude number). As you will notice, he kept in mind that the scaling laws differed between frictional and residual resistances. Finally, he added together the two individually arrived-at resistances so as to provide a reliable prediction of total full scale resistance, and from it EHP. The diagram in Fig. 8.3 illustrates Froude's logic and may help clarify the thinking outlined above.

Today, more than a century later, naval architects still exploit Froude's method. Numerous peripheral improvements have of course been developed, but the fundamental approach is still the same. Until hydrodynamicists can completely solve the basic physical laws of wave-making and the complex flow around the hull, naval architects will continue to rely on Mr. Froude's model basin techniques.

Figure 8.4 shows the 360-ft-long model basin at The University of Michigan's Department of Naval Architecture and Marine Engineering.

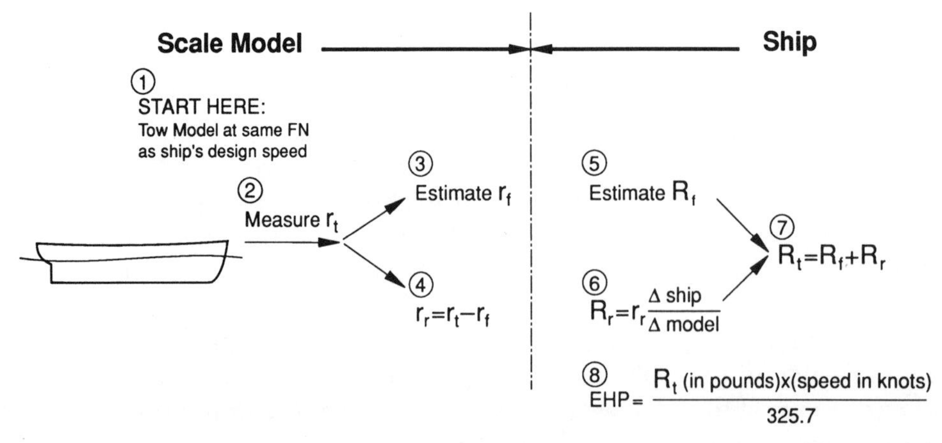

Fig. 8.3 Schematic in illustration of Froude's logical procedure for using model tests to predict a ship's effective horsepower, EHP.

Fig. 8.4 University of Michigan model basin.

Further Reading

Van Manen, J. D., and Van Oossanan, P., "Resistance," Chapter V in *Principles of Naval Architecture,* Edward V. Lewis, Ed., Society of Naval Architects and Marine Engineers, Jersey City, N.J., 1988.

Van Manen, J. D., and Van Oossanen, P., "Propulsion," Chapter VI in *Principles of Naval Architecture,* Edward V. Lewis, Ed., Society of Naval Architects and Marine Engineers, Jersey City, N.J., 1988.

Comstock, J. P., *Introduction to Naval Architecture,* Newport News Shipbuilding Company, Newport News, Va., 1944.

The Marad Systematic Series of Full-Form Ship Models, Donald P. Roseman, Ed., Society of Naval Architects and Marine Engineers, Jersey City, N.J., 1987.

Hamlin, Cyrus, *Preliminary Design of Boats and Ships,* Cornell Maritime Press, Centreville, Md., 1988.

Fig. 8.5 A selection of U.S. Coast Guard craft. The Coast Guard's multiple duties require a wide variety of floating craft. Upper left: A 110-ft patrol boat, with speed in excess of 26 knots. Upper right: A motor lifeboat. Lower left: A 180-ft seagoing buoy tender with icebreaking capability. Lower right: A 43-ft fast coastal interceptor, with speed in excess of 55 knots, used primarily for drug interdiction. Courtesy R.W. Paugh, U.S. Coast Guard.

CHAPTER IX

PROPULSION DEVICES

1. Perspective

The immediately preceding chapter addresses the topic of how a boat or ship resists motion through the water. The chapter that follows this one covers the topic of internal mechanical sources of power for overcoming that resistance. In this chapter you will learn about various methods for converting that internal source of power into the thrust required to overcome the resistance. This is the nexus between the responsibilities of the naval architect and the marine engineer. Understanding and cooperation between them are necessary to the overall design of an acceptable vessel.

Before the advent of mechanical propulsion, watercraft were moved first by human power (via paddles, oars, or poles). Humans could also pull canalboats with a rope, but they soon learned that a horse, ox, or mule was better suited to the task. Wind power dominated ocean transport until about a century ago, when steam and other forms of mechanical power came into their own. Wind power is discussed in Chapter XI. My intent in this chapter is to give you a brief introduction to mechanical propulsion devices. Before doing so, however, I must make passing mention of MHD: *magnetohydrodynamic* propulsion, a still-developing concept employing no moving parts. The system exploits the physical properties of electomagnetic fields to pump seawater through a fore-and-aft tube and so produce thrust. Current hopes for making the concept feasible hinge on the development of improved techniques for achieving superconductivity (implying extremely low temperatures). It is a development of particular interest to naval designers because it promises quiet, hence hard-to-detect, propulsion.

2. Mechanical Propulsion Devices

In today's maritime world, the screw propeller is the dominant device for imparting thrust to the hull. It is consequently appropriate to focus attention in that direction. First, however, there are alternative devices that merit at least passing mention. Paddle wheels, for example, still find a useful place in inland water excursion boats and, occasionally, on small commercial craft that operate in extremely shallow waters. Side wheels, if independently powered, offer exceptional maneuverability. The same is true of split stern wheels, called *quarter wheels.* There are many disadvantages of paddlewheels: being slow-turning, they require relatively large, heavy, and expensive propulsion machinery to provide a given amount of power; they operate well over only a limited range of drafts and are easily damaged; if at the stern, they are troubled when the ship pitches; if mounted at the sides, they are troubled when the ship rolls.

Cycloidal propellers (Fig. 9.1) turn about a vertical axis. They consist of several narrow vertical blades suspended below the hull from a flat circular rotating plate that receives its power from some prime mover such as a diesel engine. Each blade rotates on its own vertical axis as the family of blades revolves about the center of the turning plate. This rotation is controlled by internal eccentric gears in such a way that the overall system produces thrust in any desired horizontal direction, or no thrust at all when desired. With a cycloidal propeller, then, a vessel can be given extreme maneuverability without recourse to a rudder. These propellers were first commercially developed by the Voith Schneider Company (of Germany) and are therefore frequently referred to as *Voith Schneider wheels.* Gearing problems tend to limit the power that can be transmitted into a single unit, and their efficiency is not as good as that of a screw propeller. They do have a place, however, and are often used where exceptional maneuverability is desired, as in harbor tugs.

Air propellers are used on air-cushion vehicles and on small craft intended for operating in shallow, weed-infested waters where screw propellers would be unsuitable. They are noisy and relatively inefficient.

Ferry boats operating across narrow bodies of water are sometimes fitted with engines that pull the boat along on chains connected to either shore. Those chains can foul boats moving along the length of the waterway, so one must be cautious about adopting that approach.

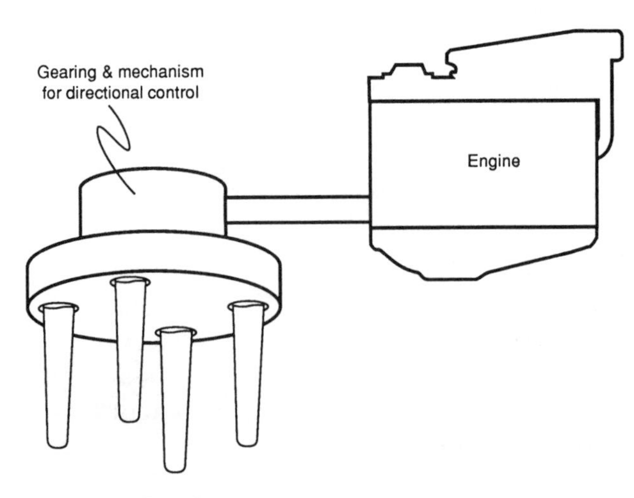

Fig. 9.1 Cycloidal propeller.

Thrust can be produced by internal pumps that discharge a jet of water over the stern. Hinged vanes at the outlet can be used to steer the boat or to reverse the direction of flow so as to make the boat stop or go astern. This kind of propulsion system works well in shallow waters where there is little room for screw propellers. Its inherent disadvantage is that a good share of the energy put into the system is wasted in overcoming friction within the ducts. In high-speed applications these water jet systems may have overall efficiencies that exceed those attainable by conventional propellers.

3. Screw Propellers

3.1 Function. A screw propeller's basic function is to convert torsional energy (as received from some internal mechanical source) into thrust. In normal applications none of the alternative propulsion systems outlined can do the job as well. The blades of the propeller are twisted in such a way that, as the unit turns, it "screws" itself through the water much as a wood screw (when twisted) pulls itself into wood. Each blade slices through the water and imparts a backward thrust as shown in Fig. 9.2. The sternward kick given the water imparts an equal and opposite thrust on the blade and the horizontal component of that, in turn, is carried through the propeller shaft to some thrust bearing within the boat.

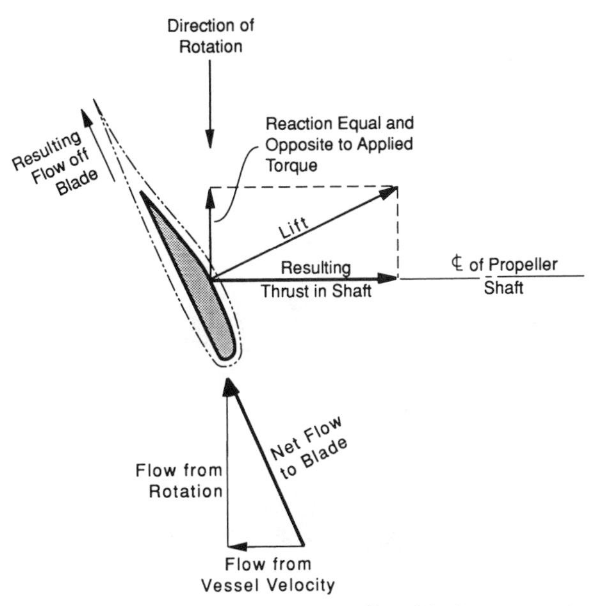

Fig. 9.2 How a screw propeller blade converts torque into thrust.

3.2 Types of screw propellers. Screw propellers are found in a wide variety of configurations. The *solid* (or *fixed-pitch*) *propeller,* as the name implies, is cast and machined in one piece. That is the kind you will find on most craft ranging from small outboard motorboats to the largest ships. A *controllable-pitch propeller* (commonly called a "CP wheel") has swiveling blades, the angle of which can be controlled from within the vessel. A complicated and expensive bit of hardware, it can be justified in vessels that operate over a wide range of loading conditions. Tugboats and trawlers are two examples. A simple variation is the *reversible-pitch propeller,* where the blades can be switched from full ahead to full astern, but nothing in between. In another variation there are two discrete ahead pitches. An *adjustable-pitch propeller,* on the other hand, has individual blades bolted to the hub. This allows the pitch of the blades to be changed if necessary; or, if a single blade is damaged, it alone can be replaced without the necessity of replacing the entire propeller.

A *supercavitating propeller* may be appropriate in high-performance craft. Ordinary propellers, if turned at high enough speed, are apt to experience *cavitation,* which you can think of as cold-water boiling. As you are probably aware, the boiling point of water drops as pressure reduces. While the front (that is, working) face of a propeller blade experiences an increase in pressure, the back face experiences a reduction. If the blade moves fast enough, the pressure may be reduced to the extent that the water will form transient bubbles that, in rapidly forming and collapsing, manage to pit and erode the blade. This condition, which is known as *cavitation,* should be avoided because the roughened surface adds to the frictional resistance and this can have serious effects on the efficiency of the propeller. A supercavitating propeller overcomes the problem by being so shaped that the phenomenon of cavitation envelops the entire suction side of the blade, but the bubbles collapse after leaving it (Fig. 9.3). These propel-

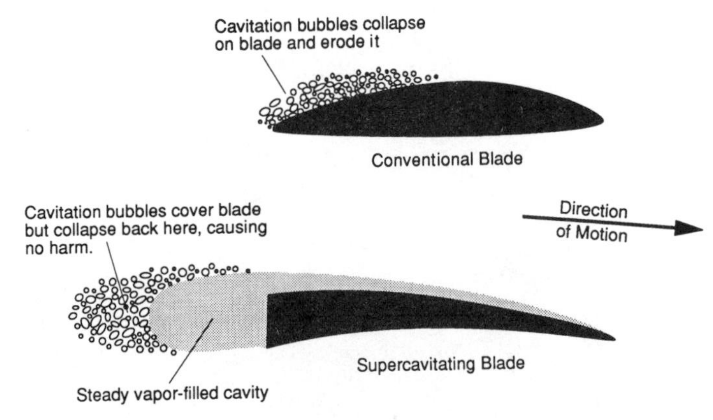

Fig. 9.3 Supercavitating propeller.

lers are inherently less efficient than ordinary propellers when operating at conventional speeds, but may be superior at high speeds. As a result their use is best confined to high-speed craft.

A *contraguide propeller* does not rotate. It is simply a twisted part of the hull or appendages (usually immediately aft of a conventional propeller) that provides some forward thrust by straightening out the spiraling rush of water leaving the conventional propeller. A more complicated way of exploiting that spiraling energy is to use *counter-rotating propellers.* Such a system comprises two propellers of about equal size turning in opposite directions on coaxial shafts. The after wheel's shaft extends through the hollow shaft of the forward wheel. Needless to say, the arrangement is too complicated and expensive to be often used except in torpedoes. The recently invented *Grimm propeller* is yet a third way to regain energy from the spiraling wake. The Grimm propeller is a free-turning unit mounted aft of and directly in line with the main propeller (its shaft being supported from aft). Its blades are longer than those of the main propeller, and their inner and outer areas are twisted in opposite directions. The inner length of each blade absorbs energy from the spiraling wake and passes it on to the outer length, which is pitched to add forward thrust to the ship.

3.3 Useful definitions. Here are a few key terms you may need to know in discussing propellers with a naval architect or hardware supplier.

A *right-hand propeller* is one that provides forward thrust when turning in the clockwise direction when viewed from astern. *A left-hand propeller* provides forward thrust when turning counter-clockwise.

The propeller *diameter* is simple enough, it being the diameter of the circle traced out by the tips of the propeller blades.

The *pitch* of a propeller is the theoretical linear distance the propeller would move straight ahead during one complete revolution if it were turning within a threaded metal nut. The *pitch angle* is the number of degrees of twist at any given point along the length of a blade. The angle varies, becoming smaller as one moves away from the hub. That allows the parts of the blade with smaller diameters to achieve the same forward advance (that is, the same pitch) despite the reduced circumferences of the circles that they sweep out in one revolution.

The *slip* is the linear difference between the pitch and the actual distance the propeller moves straight ahead through the water. Without slip the blades will have no angle of attack and there can be no useful thrust.

Skew and *rake* are not to be confused. *Skew* is observed when looking at a propeller from fore or aft. It refers to the orientation of the blades being other than radial. *Rake,* on the other hand, is observed when the propeller is viewed from the side. It refers to the condition of the blades being at less than 90 degrees to the shaft line (Fig. 9.4).

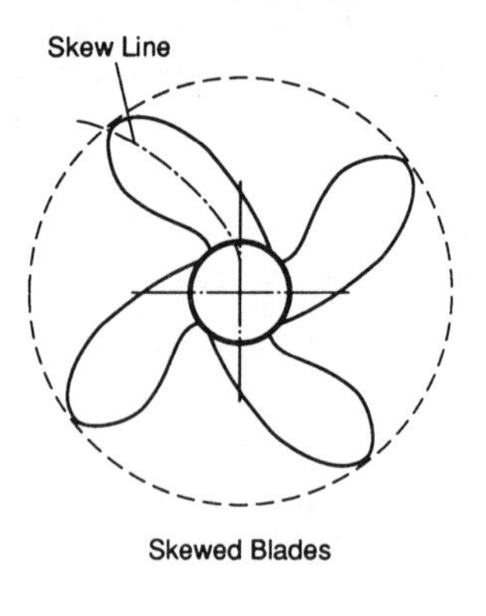

Fig. 9.4 Skewed and raked blades.

Skew and rake tend to lower efficiency a little, but both are considered helpful in reducing the vibration that may be brought on by an uneven flow of water into the propeller. (Owing to the shape of the typical hull, the wake is likely to be more pronounced near the water's surface.) A propeller with an extreme degree of skew may also be called a *weedless propeller* because it is less likely to tangle itself in weeds.

A *square propeller* is one in which the pitch equals the diameter.

A *duct* (or *Kort nozzle*) is a circular structural shroud surrounding a propeller. When properly designed, such a nozzle increases the thrust of the propeller. They are most effective in towboats or tugs, where high thrusts and low speeds reign. In ships of normal or higher speed, the added frictional resistance of the nozzle is likely to cancel the benefits. A variation is the *steering nozzle,* which allows the shroud to serve as a rudder as well as a thrust-enhancer.

A *propeller tunnel* is formed when the bottom of a vessel is scooped out so as to allow a large (hence efficient) propeller to be fitted to a shallow-draft vessel (Fig. 9.5). These are common features on river towboats. As you may well imagine, they tie in nicely with propeller ducts.

The *tailshaft* is the aftermost part of the propeller shaft and is keyed into the propeller hub. Where it passes through the afterpeak bulkhead it is encircled by self-lubricating soft material that is squeezed against the shaft to minimize entry of seawater. This fitting is called the *stern gland,* or shaft seal.

A *line shaft* is the straight length of propeller shaft between the engine, or gears, and the tailshaft. If very long, it will be made up of several sections all bolted together. It may be carried in a narrow *shaft tunnel* (or *shaft alley*), which protects it and allows the crew to lubricate the bearings and attend to the stern gland.

3.4 Transmissions. The previous paragraph implies that the propeller shaft is led straight aft from the engine. That is not always the case. In the familiar outboard motor, for example, the power is transmitted through an L-shaped configuration of shafting and gears. That saves room and allows the engine to steer as well as drive the boat. In some speedboats, the common inboard/outboard arrangement employs a Z-shaped configuration. On some small commercial vessels, diesel engines mounted on the after deck are also attached to the propellers through a Z drive. If you travel to Thailand, you will see riverboats, each powered with an automobile engine mounted on a swiveling bed at the stern. Power goes directly into a long, unsupported shaft that runs diagonally down and aft into the water with the propeller at the lower end. These are called *long-tail boats.* The engine can be used to steer the boat and, when tangled in weeds or plastic bags, they can be tilted up so someone in another boat can reach over and clear the propeller.

For vessels driven by steam turbines, gasoline engines, or high-speed diesels, the efficient rotational speed of the machinery is far greater than can be accommodated by an efficient screw propeller. In such cases, reduction gears are usually used in order to bring shaft speeds down to practical values.

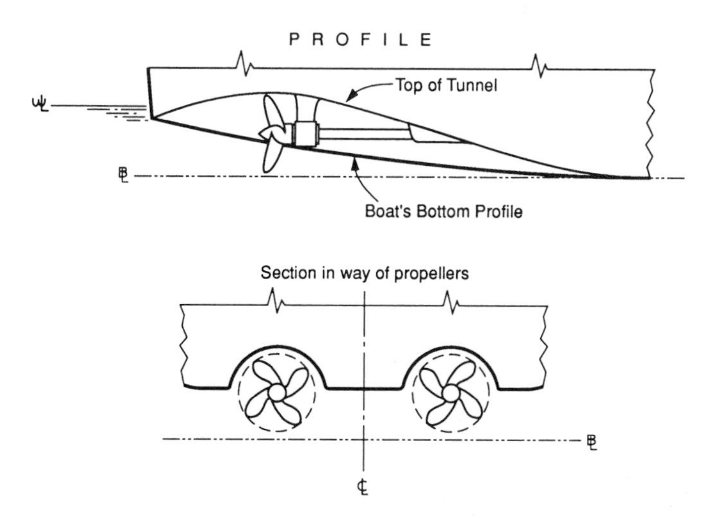

Fig. 9.5 Typical tunnel stern.

In ships with electrical transmission systems, the driving motor can be located in the stern just ahead of the propeller, while the prime mover (that is, the engine) may be in some other, perhaps distant, location. Although electrical transmission suffers from low efficiency, there are many types of vessels where it may be used to advantage. In icebreakers, for example, the motors serve as capable shock absorbers that protect the engine from the sudden, potentially damaging, changes in torque that arise as the propeller encounters large chunks of ice. As superconducting electrical systems become more practical, they will lead to new freedoms in ship arrangements.

The subject of transmissions is touched on again in the next chapter.

3.5 Design objectives. The central aim in designing (or selecting) a screw propeller is to effect a proper match with the propulsion machinery. The propeller must provide the thrust required to move the vessel through the water at the design speed while revolving at a rate appropriate both to itself and to efficient operation of the engine. At the same time the torque going into the propeller must impose proper cylinder pressures in the engine (assuming the engine to be diesel or gasoline). If a controllable-pitch propeller is selected, this task is greatly eased. At the other extreme, where a fixed-pitch propeller is directly attached without reduction gear to a diesel engine, some compromise is often required. In many such arrangements, the rpm is generally higher than would be desirable for the propeller and lower than would be desirable for the engine. Like most other matters in marine design, the most efficient overall system is often made up of less-than-optimum components.

The delicate balancing act described above must be done within certain constraints. First of all, there is the matter of locating the propeller so as to minimize the probability of its being damaged. Except in the case of outboard motors, the bottom of the propeller should not extend below the vessel's baseline. (There are exceptions to this in boats or ships where high speed is of the essence: racing motorboats, destroyers, and so forth.) In multiple-screw vessels the propellers are susceptible to striking piers, lock walls, buoys, and tugboats. For that reason they should be kept as far inboard as possible. Where such propellers are thought to be in danger they may be protected with a heavy framework, called a *propeller guard,* protruding from the hull above the propeller.

Another constraint is imposed by the necessity of avoiding cavitation. This will influence such questions as propeller blade tip speed (a function of diameter and rpm), blade area, and thrust loading.

A third constraint is that of avoiding propeller (or propeller induced) vibration. What is of chief importance here is to assure a smooth, as-uniform-as-possible flow of water into the propeller. The first rule is to provide generous clearance between hull and propeller.

The hull stern lines should be configured as much as possible to give equality of wake throughout the propeller disk.

A fourth constraint is that of resistance to structural failure. For any given material, certain thicknesses of blade will be required. This must not be overdone, however, because thicker blades lead to loss of efficiency.

Another practical consideration is that of arranging the tailshaft so it can be withdrawn from the stern tube for periodic inspection. Most tailshafts are so arranged that they can be pulled into the hull. In other cases the tailshaft must be drawn outboard, possibly causing a problem of interference with the rudder. In either scheme, the propeller must first be removed from the tailshaft.

3.6 Design variables. Designing a screw propeller requires many decisions. First of all, one must settle on the type (as outlined in Section 3.2). Assuming the choice to be a solid-screw propeller, the next thing is to decide whether to use single or multiple units. Single screws are the least expensive and generally most efficient. They are therefore used whenever possible. In the case of high-speed ships, or ships requiring greater-than-average power, reliability, or maneuverability, multiple screws may be in order.

In most cases the diameter should be made as great as possible within the constraints described in Section 3.5. The number of blades deserves careful thought. In small craft, two to four blades are usually found appropriate; in large commercial ships, they usually number from four to seven. As the number of blades increases, the width of each blade can be reduced, which may increase its efficiency. Larger numbers of blades also frequently minimize vibrations. On the other hand, larger numbers lead to mutual interference between the blades, which may cancel the gain in efficiency brought on by the narrower blades. As you may infer, the questions of propeller diameter, number of blades, and area per blade are all intimately related.

Another important variable is the pitch of the blades. To have the correct angle of attack at a given vessel speed, a slow-turning propeller will require a steep pitch, while a fast-turning propeller can get by with a smaller pitch. At the extremes of the speed range, fast motorboats require relatively small-diameter, high-speed propellers with high pitch, while slow tugboats require large-diameter, slow-turning wheels. These are important parameters that must be balanced in achieving a proper integration of propeller and engine.

As mentioned in Section 3.5, the designer must not forget to make the propeller strong enough to stand up in service. Special attention must be paid if the vessel is intended for operating in ice.

Finally, the designer may choose to introduce at least moderate amounts of rake and skew. These may help reduce vibrations but only at some modest loss in efficiency.

3.7 Design techniques. For outboard motor boats and similar small craft, there is little virtue in essaying a careful propeller design study. Put a typical load aboard your boat and then spend a couple of hours trying out stock propellers with various combinations of diameter and pitch. Do this only with the counsel of some experienced mechanic so that you do not overload your engine or allow excessive piston speeds.

For larger vessels, where tailor-made propellers are in order, engineers can make use of standard designs adapted to any particular set of conditions. Designers are aided in this by propeller design charts derived from large numbers of model tests for well-designed propellers. The models are methodically varied in each of the key parameters: pitch ratio, speed of advance, number of blades, power, and so forth. The tests measure the efficiency of each model under each set of circumstances. These are called *open-water efficiencies* because the tests are carried out with water flowing to the propeller uninhibited by any hydrodynamic diversion from hull, rudder, or other appendages.

Figure 9.6 shows a sample section of a typical propeller design chart. The horizontal scale measures a coefficient identified here as "B," which recognizes the propeller's rpm, delivered power, and speed of advance through the water. (I am not going to burden you with the

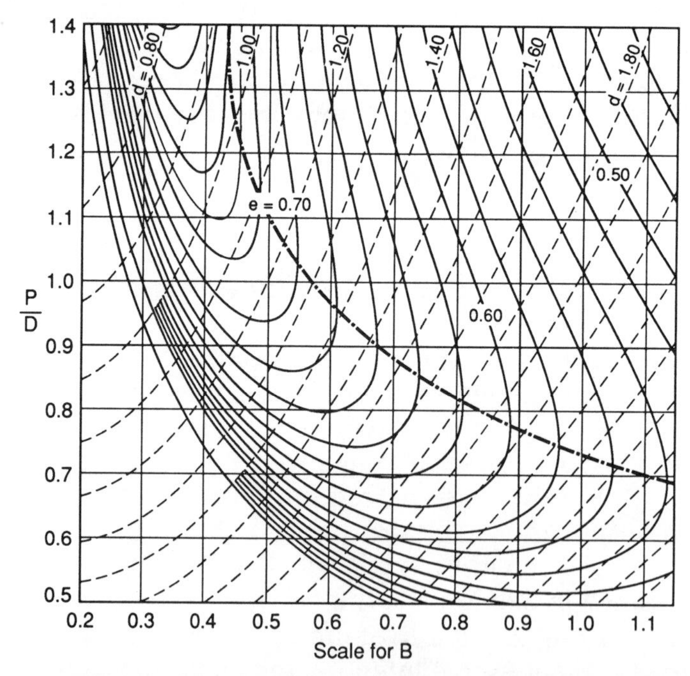

Fig. 9.6 Sample section of a typical propeller design chart. Solid lines labeled "e" are contours of equal open-water efficiency. Dotted lines labeled "d" are contours of equal advance coefficient. The dot-dash curve is the locus of maximum efficiencies.

details.) The vertical scale shows the ratio of pitch to diameter. There are two families of contours; one shows open-water efficiencies, the other values of the advance coefficient. The latter equals the rpm times the diameter divided by the speed of advance. There are charts available for different numbers of blades and different values of *blade area ratio* (the total area of the blades divided by the area of the propeller disk).

The designer first selects the appropriate chart (or more likely, selects two or more charts and interpolates between them). Knowing the values entering into the "B" scale, and those entering into the advance coefficient ("d" contours) allows one to read off the corresponding open-water efficiency ("e" contours) and pitch ratio (shown on the vertical scale). By systematically varying the several physical characteristics of the design, the designer eventually finds the combination of parameters promising best possible efficiency.

You can be sure there are many ramifications left unsaid in the paragraphs above. I think, however, that what I have laid before you will give you an appreciation of what is behind the science of propeller design.

Further Reading

Van Manen, J. D. and Van Oossanen, P., "Propulsion," Chapter VI in *Principles of Naval Architecture,* Edward V. Lewis, Ed., Society of Naval Architects and Marine Engineers, Jersey City, N.J., 1988.

Barnaby, K. C., *Basic Naval Architecture,* Hutchinson's, London, 1948.

Hamlin, Cyrus, *Preliminary Design of Boats and Ships,* Cornell Maritime Press, Centreville, Md., 1989.

Fig. 9.7 Stern frame, rudder, and highly skewed propeller. The propeller blades are skewed as a means of reducing wake-induced vibrations. The semibalanced rudder is supported by an appendage (horn) cast integral with the stern frame. Courtesy Peter E. Jaquith, National Steel and Shipbuilding.

Fig. 9.8 Propeller tip vortex cavitation. This stroboscopic photograph (made in a propeller tunnel) shows clearly how the water leaves the propeller in a neat spiral. Courtesy J. D. van Manen and P. van Oossanen, Netherlands Maritime Research Institute.

CHAPTER X

CHOOSING PROPULSION MACHINERY

1. Perspective

My intent in this chapter is to present a brief introduction to a large and complex subject: that of making wise choices from among the many available combinations of propulsion prime movers, propellers, transmission systems, and associated auxiliary machinery. Naval architects and marine engineers must not only understand the advantages and disadvantages of each combination, but also must recognize their influence on the overall arrangement and economics of the ship or boat they are called upon to propel.

Integral with the above is the question of selecting appropriate fuels, which also is covered in this chapter.

2. Alternative Systems

2.1 Single versus multiple screws. The majority of mechanically propelled boats and ships in this world have but a single screw. The reasons for this are many; principal among these are simplicity, ease of engine control, and maximum propulsive efficiency. These technical factors lead to appreciable savings in costs of construction and operation. Multiple screws are nevertheless desirable or necessary under certain sets of conditions. In extremely high-powered ships, for example, practical limits (principally, danger of propeller cavitation) on the power to be extracted from a single propeller may dictate more than one. Twin screws enhance maneuverability; pushing ahead on one and pulling astern on the other helps the vessel turn in confined waters. If one engine fails, or is damaged in battle, the other is still available to propel the ship. In shallow-draft vessels, where propeller diameters are severely limited, efficient propulsion dictates multiple screws because a single screw would be ill-suited to convert the required power into usable thrust. That would lead to low efficiency and, perhaps, problems with blade cavitation or vibrations, or both.

The number of propellers and the number of propulsors need not be the same. Two or more prime movers can be coupled to a single shaft either mechanically or electrically.

2.2 Prime movers. Almost all mechanically propelled boats and ships use either gasoline engines, diesel engines, gas turbines, or steam turbines. In a minority of cases, two of those types may be combined. Steam-reciprocating engines are now largely historic relics, and marine steam turbines (aside from those associated with nuclear reactors) appear to be headed in the same direction.

Gasoline engines, diesel engines, and gas turbines are all what we call internal-combustion engines; by this we mean that the fuel is burned within the engine. Steam turbines and steam engines, on the other hand, burn the fuel in a separate device, namely a boiler, or use steam generated in a nuclear reactor.

In a gasoline engine, power is derived from a combustible mixture of vaporized gasoline and air. Introduced into the cylinders and compressed, this mixture is ignited by a spark. A diesel engine is similar except that the fuel is injected directly into the cylinder and ignition is triggered by the heat of compression; spark plugs and carburetors are not required. The greater pressures involved require stronger engine blocks so, of the two, diesel engines are considerably the heavier.

You may hear mention of "two-stroke" versus "four-stroke" engines. In a *two-stroke* gasoline engine, ignition occurs every time a piston reaches the top of its stroke. Energy is transferred into the crankshaft during part of the downward stroke. During the rest of the downward stroke and start of the upward stroke the exhaust gases are withdrawn from one part of the cylinder. During the upward stroke, a mixture of air and gasoline (from the carburetor) is forced into another part. As the piston continues its upward stroke, the intake and exhaust ports are closed and the combustible mixture is compressed. At the top of the stroke, the spark plug ignites the mixture and the process is repeated.

In a *four-stroke* gasoline engine, ignition occurs every other time a piston reaches the top of its stroke. After ignition is initiated the piston transmits energy to the crankshaft during its entire downward stroke. During its first return stroke, an exhaust valve at the top of the cylinder is open and the exhaust gases are forced out. During the piston's second downward stroke, the exhaust valve is closed; at the same time, an inlet valve is open so that the descending piston draws in the combustible mixture of air and gasoline. The mixture is then compressed as the inlet valve is closed and the piston returns to the top of the cylinder to complete the four-stroke cycle.

Two-stroke engines are smaller and lighter than four-stroke engines, but also less fuel-efficient.

In diesel engines, the two-stroke and four-stroke principles are the same as outlined above. The main difference is that the fuel is sprayed into the cylinder near the end of the compression stroke rather than being first mixed with air in a carburetor.

There are three main categories of diesel engines based on speed of rotation. Slow-speed engines normally turn from 75 to 250 rpm.

Designed and built specifically for ship propulsion, they are connected directly to the propeller. In general, medium-speed diesels are adaptations of stationary power plant or locomotive engines. They range in speed between 400 and 900 rpm. Most high-speed diesels are adapted from truck engines and range in speed between 1000 and 2400 rpm. High-speed diesel propulsion systems are used principally in smaller vessels where compactness and weight saving are of overriding concern (see Woodward, 1981).

A gas turbine derives its energy from internally expanding gases of combustion flowing through its fan-like blades. A steam turbine is much the same except that what flows through the blades is steam from an external boiler. Gas turbines, having no need of a boiler, are extremely compact. On the other hand, their higher internal temperatures require much more expensive materials of construction. Both kinds of turbines rotate at such high speed as to require double reduction gears to bring the shaft speed down to what is appropriate for an efficient propeller.

There are two major classes of marine gas turbines: those derived from aircraft engines and those derived from stationary power plants. The aircraft type turbines are lighter and smaller, but suffer from higher fuel costs and shorter life.

Combatant naval vessels have two widely differing modes of operation while at sea. Most of the time they operate at "cruising" speed during which time fuel economy is important. On occasion, however, they need to go at top ("flank") speed. To meet these changed demands they are often fitted with a main unit (steam turbine or diesel) sized to provide cruising speed and with a supplementary unit that can be brought in on top to produce flank speed. Gas turbines are popular for this supplementary service because they are compact. Moreover, their relatively poor fuel rate is no great disadvantage because they are seldom called upon. The most common such combination is called CODAG, an acronym for Combined Diesel And Gas (turbine).

2.3 Transmissions. Transmission systems have the important function of connecting the prime mover to the propeller and passing the propeller thrust into the hull. In the simplest case, the direct-connected, slow-speed diesel engine, the transmission system is just a straight shaft carrying torque to the propeller without change in speed of rotation. In other cases, however, the system may be called upon to reverse the propeller's direction of rotation without changing that of the prime mover. Most steam turbines can be reversed, but gasoline engines and gas turbines usually cannot. Some diesel engines can be reversed, others cannot.

Transmissions may contain reduction gears so as to effect compatibility between the rpm of the propeller and that of the prime mover. Propellers work most efficiently at relatively low speeds, seldom over 100 rpm on large ships. Most engines, on the other hand, can be made smaller if allowed to work at higher speeds. *Single-reduction* gears allow the engine to turn perhaps four to seven times as fast as the propeller. They are most often used with medium-speed diesel engines. *Double-reduction* gears allow the engine to turn from ten to fifty times as fast as the propeller. They are appropriately used with high-speed diesels, gas turbines, or steam turbines.

An altogether different approach to transmitting energy from the prime mover to the propeller employs an electrical system. The prime mover drives an AC or DC generator. The derived energy is carried through heavy electrical cables to a driving motor, possibly geared down in speed of rotation, which turns the propeller. Electrical systems lose about 8 to 15 percent of the energy being transmitted and are therefore not as popular as mechanical systems, which lose only a few percent. They do, however, offer the advantage of greater versatility in layout and good maneuverability; they may eliminate the need for a shaft tunnel and they provide good shock-absorbing properties that protect the gearing and main engine from damage (especially important in icebreakers).

2.4 Alternative combinations. Figure 10.1 shows the most commonly used combinations of propulsor, transmission system, and propeller. The CODAG system has already been mentioned, but other less often used combinations are also possible.

2.5 Key considerations. In selecting a propulsion system, the designer must balance out a good many influential, often conflicting, factors. Table 10.1 lists representative values for four of the most important considerations: required investment (that is, first cost), overall weight, overall length, and fuel rate.

2.6 Overall considerations. The factors shown in Table 10.1 are by no means the only ones to consider when selecting components for a propulsion system. Marine engineers must also weigh the relative values of reliability, economic life, freedom from likelihood of explosion (always a threat in gasoline engines), crew numbers and skills, fuel costs, fuel availability, and fuel treatment requirements. They must also think about maintenance and repair requirements, availability of replacement parts, ease of access, ease of control, ease of reversing and other matters of maneuverability, reversing power, and adaptability to specific hull configurations. Finally, the list concludes with lubricating oil needs, ability to operate at fractional power without stalling, shock resistance, and relative freedom from vibration, noise, heat, odors and threat of air and water pollution.

Not to be forgotten is the practical matter of engine output powers that may be commercially available. There are certain exceptions to these numbers, but the most readily acquired units are usually found within the power ranges indicated in Table 10.2.

In addition, any particular shipowner's preferences will be heavily influenced by past experience. Owner's opinions and prejudices, whether justified or not, must always be considered.

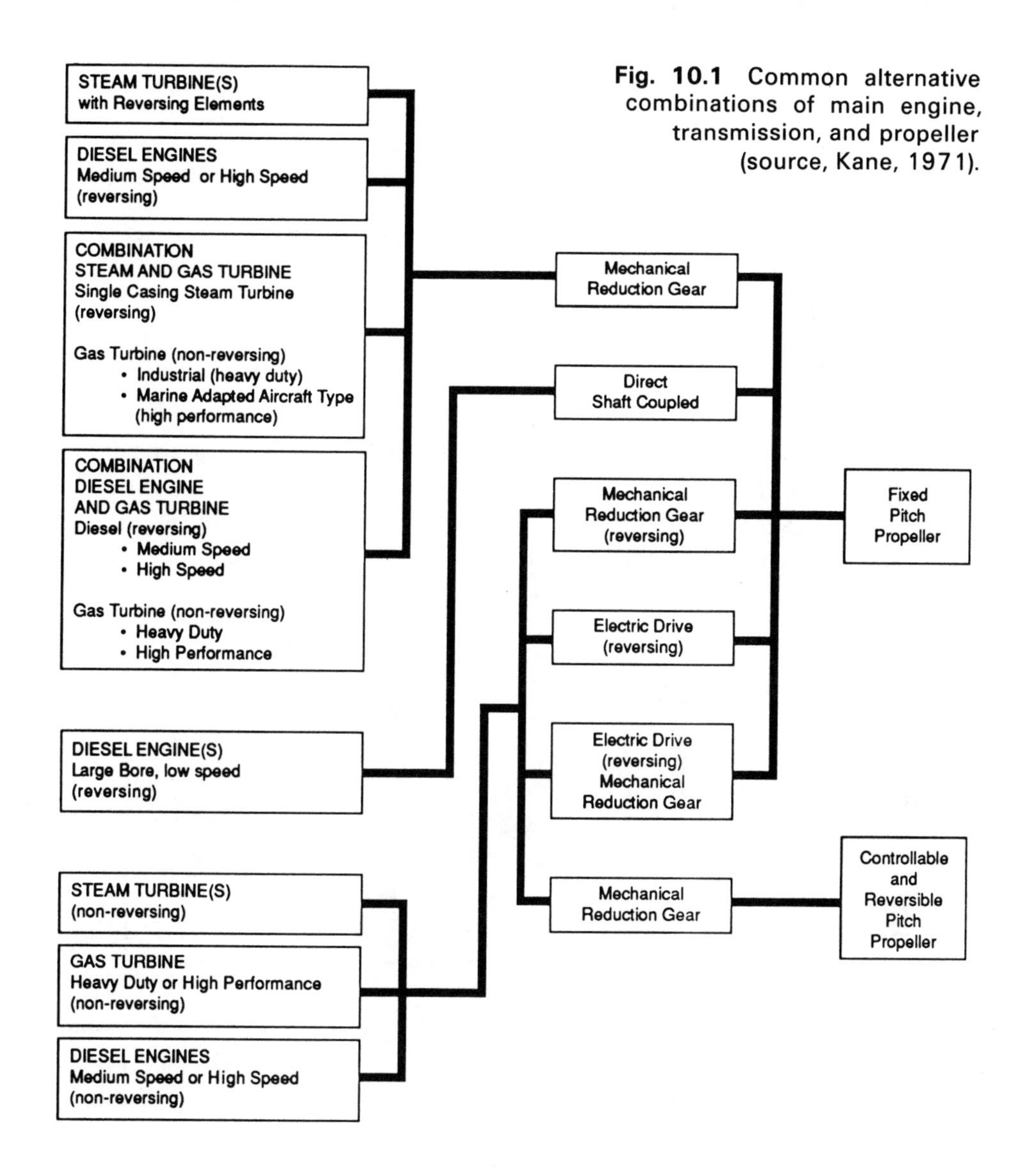

Fig. 10.1 Common alternative combinations of main engine, transmission, and propeller (source, Kane, 1971).

Table 10.1 Relative virtues of four kinds of marine propulsors

Type of Machinery	Relative First Cost	Relative Overall Weight[a]	Overall Length, ft[b]	Fuel Rate[c]
Direct drive diesel	1.00	1.00	101	0.38
Steam turbine[d]	0.78	0.94	57.5	0.49
Gas turbine[d]	0.80	0.83	57.5	0.43
Medium-speed diesel[d]	0.48	0.81	57.5	0.40

[a] Relative overall weights are based on 16 000 SHP plants with fuel for a 10 000-mile voyage.

[b] Overall lengths are appropriate to 16 000 SHP plants.

[c] The fuel rates include energy for electrical generation. The numbers shown are in pounds of fuel per horsepower-hour.

[d] Geared.

Table 10.2 Usual horsepower ranges

Outboard motors	less than 1 to 250
Gasoline engines	10 to 250
High-speed diesels	90 to 2000
Medium-speed diesels	800 to 30 000
Low-speed diesels	800 to 60 000
Gas turbines	up to 50 000 per shaft
Steam turbines	up to 70 000 per shaft

As a practical illustration, Fig. 10.2 shows the arrangements of the main machinery units in an oceanographic research ship. This particular vessel uses three high-speed diesel engines to supply electrical energy to two motors that are connected to a single propeller shaft.

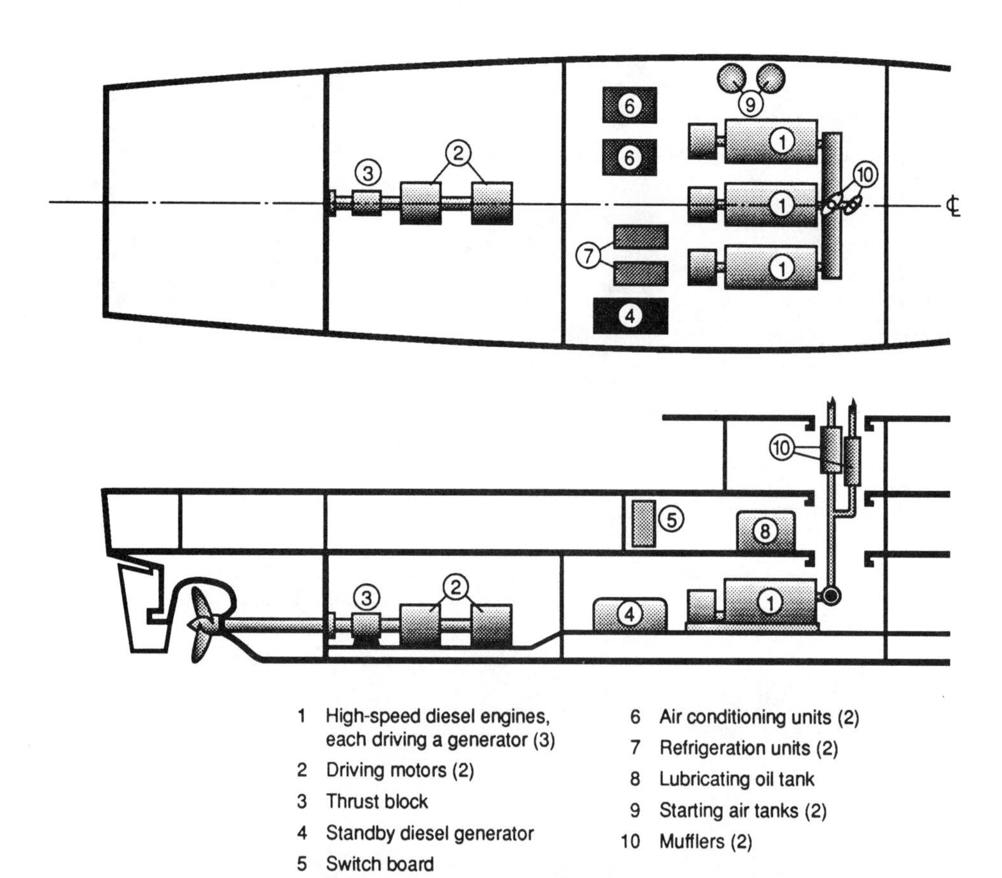

Fig. 10.2 Machinery arrangements in an oceanographic research vessel. In this case you see a diesel-electric propulsion system with three diesel generators side-by-side and two propulsion motors in tandem driving a single screw. Each of the three main diesel engines is capable of producing about 1000 HP.

3. *Generating Electricity*

Over the years we have seen an ever-increasing need for electrical energy on shipboard. The introduction of auxiliary maneuvering devices, such as bow or stern thrusters, and special cargo-handling or cargo access gear are prime examples. Moreover, experience shows that during the useful life of a ship or boat the owner is likely to add devices that will place extra demands on the generators. These can range from a little refrigerator in some room to the addition of an air-conditioning system for the entire accommodation space. I recall being on an older ship in which the generators were overloaded and showing clear signs of overheating. To cool the generator, the engineers were using an electric fan to blow air at it. The question then arose, was the added wattage of the fan warming the generator more than the air was cooling it? The moral of all of this is that the prudent owner will specify some generous margin in any new vessel's generating capacity.

In a matter of terminology, electrical energy needs are divided into two parts: engine room and ship's service requirements. The engine room energy is needed, for example, to drive the many motors for pumps associated with the main and auxiliary machinery. The ship's service category includes all the power requirements to make the accommodations habitable (what we call the *hotel load*) and to operate navigating equipment, cargo gear, mooring gear, bilge and ballast pumps, steering engine, mechanical doors, and many other items within the realm of hull engineering. Such items of equipment are seldom, if ever, all in operation at the same time, so the designer must visualize the vessel in service and try to establish the likely set of circumstances that will place peak demand on the generating system.

A ship of any size will probably carry at least two independent generating units for normal use. These will be so sized that either can be shut down for repairs without seriously impairing operations of the ship. In addition, most regulatory agencies require the presence of an emergency diesel generator that must be located above the freeboard deck and outside the engine room or its overhead casing.

In diesel-powered ships the most fuel-efficient way to produce electricity is to fit a generator integral with the main engine drive shaft. This is most practical when associated with a controllable-pitch propeller, otherwise one will need to introduce complications to overcome difficulties that arise whenever engine speeds are reduced for one reason or another.

In addition to shaft-driven generators, electrical energy may also be derived from diesel-powered generators. Most of these burn a relatively high grade of fuel, whereas the main engine burns something considerably cheaper. Alternatively, generators may be powered by steam turbines, with the steam being supplied from boilers heated by exhaust gases from the main engine, whether diesel or gas turbine. In steamships, the steam supply may be taken directly from the main boilers. As an alternative, the steam may be taken from some stage of the cycle as the steam does its work while moving through the turbines. Where electrical transmission systems are employed, most of the electrical energy for auxiliary machinery and ship service can be drawn directly from the main generator.

4. *Fuels*

4.1 Alternatives. Mechanically propelled ships or boats may obtain their energy from a wide variety of sources: petroleum products (derived from crude oil), coal, nuclear fission (or, perhaps in the far future, fusion), gas (perhaps in the form of boil-off from a liquefied-gas cargo), wood or wood chips, mixtures of powdered coal and petroleum, and liquid fuel derived from coal.

However, since the great majority of today's vessels, large or small, are powered by petroleum products, the rest of this section will be confined to that sector.

4.2 Fuel categories. Petroleum fuel is most usefully categorized according to the type of engine for which it is best suited. These include principally: gasoline engines, gas turbines, high-speed diesels, medium-speed diesels, low-speed diesels (the last two are sometimes lumped), and boilers.

Fuels are also categorized according to their important physical properties. These are briefly outlined below.

Heating value—This recognizes the amount of heat energy available in each unit of the fuel. It is measured in British Thermal Units (Btus) per pound, or in kilojoules per kilogram (kJ/kg).

Ignition delay—Another related, important characteristic is the ignition period. This is identified by what is called the *Cetane rating,* the normal value of which is 100. A high number indicates a short delay, which is most critical in high-speed diesel engines.

Density—The usual measure of this characteristic is the number of *API degrees.* (API stands for American Petroleum Institute). This characteristic is indicative of the ease of separating out any water that may have found its way into the fuel. This is more easily carried out with low-density fuels because high-density fuels have a specific gravity close to that of water. Heat may be required to lower the fuel's density so as to speed the separation process.

Fluidity (pour point and viscosity)—The *pour point* is simply the temperature at which the fuel's viscosity becomes low enough to permit the material to be poured. (Many marine fuels must be heated before pumping.)

Safety: flash point—The *flash point* is a measure of volatility. It is the minimum temperature at which vapors of the material will ignite. Minimum acceptable values for safe operations are usually set at 140 to 150°F.

Cleanliness (purity)—This measures the amounts of water, ash, and sediment in the fuel.

Corrosiveness—This measures the amounts of undesirable chemical impurities such as sulfur, vanadium, sodium, potassium, calcium, and lead.

4.3 Grades of diesel oil. A commonly used standard for grading diesel fuel is that established by ASTM (American Society for Testing and Materials). The three principal grades are as follows.

No. 1-D—A low-viscosity distillate fuel oil for engines in service requiring frequent speed and load changes.

No. 2-D—A distillate fuel of lower viscosity for engines in industrial and heavy mobile service. This is the standard diesel fuel used in cars and trucks.

No. 4-D—A fuel for low- and medium-speed engines.

One of the most common designations of petroleum fuels is that of the viscosity as measured on the *Redwood Number 1 Scale.* This is an arbitrary scale citing the number of seconds required for a given amount of oil to flow through a given orifice when heated to 100°F.

4.4 Common names for petroleum fuels. The marine industry is plagued with a plethora of semantical confusions with respect to the designation of various kinds of petroleum fuels. What one operator calls "marine diesel medium," for example, another may call "gas oil." Habermann (1971) presents a table listing the many common names for ordinary petroleum fuels. One is led to the inescapable conclusion that, in discussing questions of fuel, one is well advised to define terms from the start of the conversation.

4.5 Operating profiles. How much fuel should be carried in order to complete a given voyage? A rational answer to this question first requires a careful analysis of the levels of power needed during each of the segments of the voyage. Speed constraints in harbors, rivers, or canals will dictate one level of power (and rate of burning fuel). Queuing delays will dictate another. Cargo-handling requirements will call for yet another. Normal-speed operations in open waters will be the principal component of this analysis, but one must recognize that fog or heavy seas may distort the picture.

Having figured all of the above, the analyst must then estimate the amount of fuel needed during each segment of the voyage. This requires consideration of the hours involved at each power level and the rate at which fuel is burned. Next, all those incremental amounts are added to arrive at the total fuel required for the given voyage. Then the operator must decide whether to take on bunkers (that is, fuel) at each end of a round-trip voyage or at only one. Finally, to be at all prudent, the operator must add an appreciable margin for contingencies.

In the most careful studies, the analyst figures not only the fuel used in the main engine(s) but also that used in the generators.

The analytical procedure outlined above was first developed during the days of steam engines and was called the *steaming profile.* Although the ship may be powered by diesel machinery, the analysis may still be referred to as the *steaming profile.* (The same historic influence is found in many of today's "steamship" companies whose fleets are entirely diesel-propelled.)

References and Further Reading

Woodward, John B., *Low Speed Marine Diesel,* John Wiley & Sons, New York, 1981.

Kane, J. R., "General Considerations in Marine Engineering," Chapter I in *Marine Engineering,* Roy L. Harrington, Ed., Society of Naval Architects and Marine Engineers, Jersey City, N.J., 1971.

Habermann, C. E., "Petroleum Fuels," Chapter XXIII in *Marine Engineering,* Roy L. Harrington, Ed., Society of Naval Architects and Marine Engineers, Jersey City, N.J., 1971.

Hamlin, Cyrus, *Preliminary Design of Boats and Ships,* Cornell Maritime Press, Centreville, Md., 1989.

Fig. 10.3 Modern ship propulsion. Upper: A low-speed diesel engine under test on the shop floor. Courtesy E. Jung, Sulzer Diesel U.S. Inc. Lower: Engine room control compartment. Courtesy Arthur J. Haskell, Matson Navigation Company; photo by Fisher Photo.

Fig. 10.4 Ship propulsion in a bygone era. This late 19th-century photograph shows a paddle-propelled river ferry powered by four mules on treadmills (barely visible under the shelter). The shovel allows handy disposal of spent fuel. Source unknown.

CHAPTER XI

WIND POWER

1. *Perspective*

Sailing craft, whether large or small, seem to hold an almost mystical attraction to most people. There are those who enjoy the challenge of racing. Others take pleasure in simply loafing around in a comfortable day sailer. Still others seek adventure in extended cruising to distant shores. Surely every book on naval architecture deserves a chapter on the art and science of wind propulsion. In truth, the subject merits an entire book, so what follows is but a brief summary of some of the more important aspects of sailing-craft design.

Designing an efficient sailing craft is a fine art. It requires a thorough understanding of the complex interrelationship between the forces of wind and water—aerodynamics and hydrodynamics, to use the proper terms. But, that is not enough. The well-designed sailing vessel effects a nice balance between those forces despite continually changing external and internal circumstances. These changes include wide variations in wind strength, wind direction, vessel speed, degree of leeway, and angle of heel. Other changes include various combinations of sails deployed and their setting relative to the wind, degree and location of on-board weights, rudder shape and location, and hull profile variations made possible with movable devices such as centerboards. What all this means is that the well-designed sailing craft usually arises from the mind of a naval architect who understands in a qualitative way the aerodynamics and hydrodynamics at play and who attains an ideal balance between them largely through intuitive judgment based on experience. The successful designer must be intimately acquainted with the physical conditions of the waters in which the boat or ship is meant to operate. Clearly, too, he or she must emphasize different constraints when designing racing yachts, cruising yachts, or commercial sailing ships.

In this chapter I intend to clarify the principles of how the wind and water forces are brought into balance. I say a few words about sailing yacht hull forms, keels, and rudders. You will find sketches illustrating several rigs appropriate to sailing yachts. Finally, there is a brief commentary on the application of wind propulsion to commercial ships.

2. *Balancing Forces*

2.1 Apparent wind. As a sailboat moves ahead, its forward motion will affect the speed (hence force) and direction of the apparent wind striking the sails. This is shown in Fig. 11.1. The drawing shows a true wind blowing at right angles to the boat's course ("off the starboard beam"). However, because of the boat's forward motion, the wind appears to come from over the starboard bow. If you were on the boat, the wind would seem to come over the starboard bow, and the sails should be trimmed accordingly. In low-resistance craft, such as catamarans or iceboats, this effect is so pronounced that boat speed can easily exceed true wind speeds. In light airs and calm waters, even normal displacement-type boats may enjoy this phenomenon. If the boat is headed downwind, no such effect is possible because then the boat speed subtracts from wind speed.

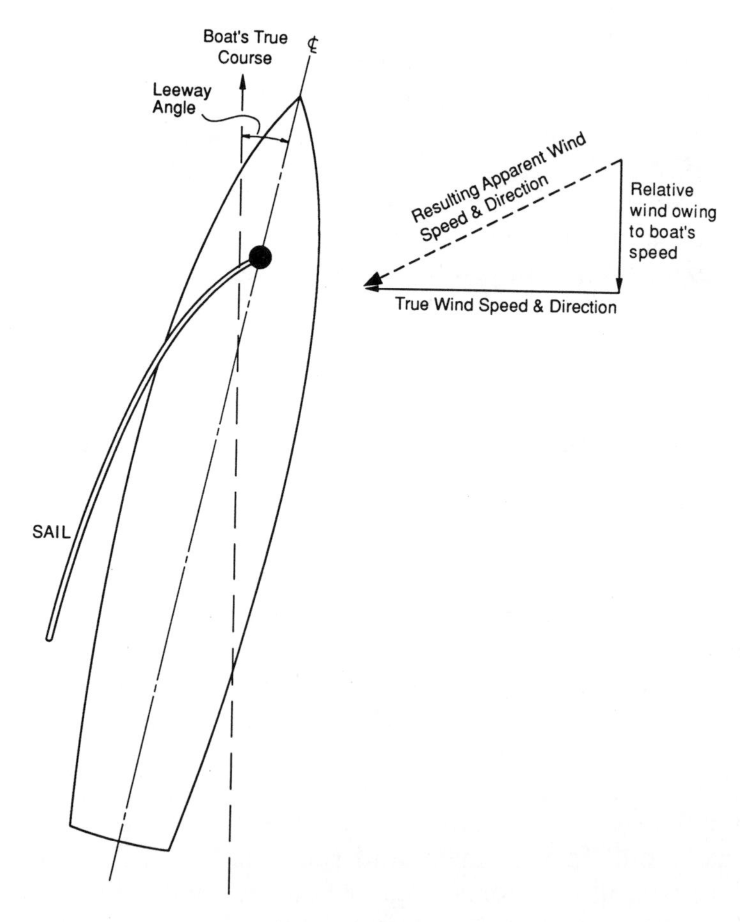

Fig. 11.1 How a boat's forward speed affects the apparent wind speed and direction as it impinges on the sail.

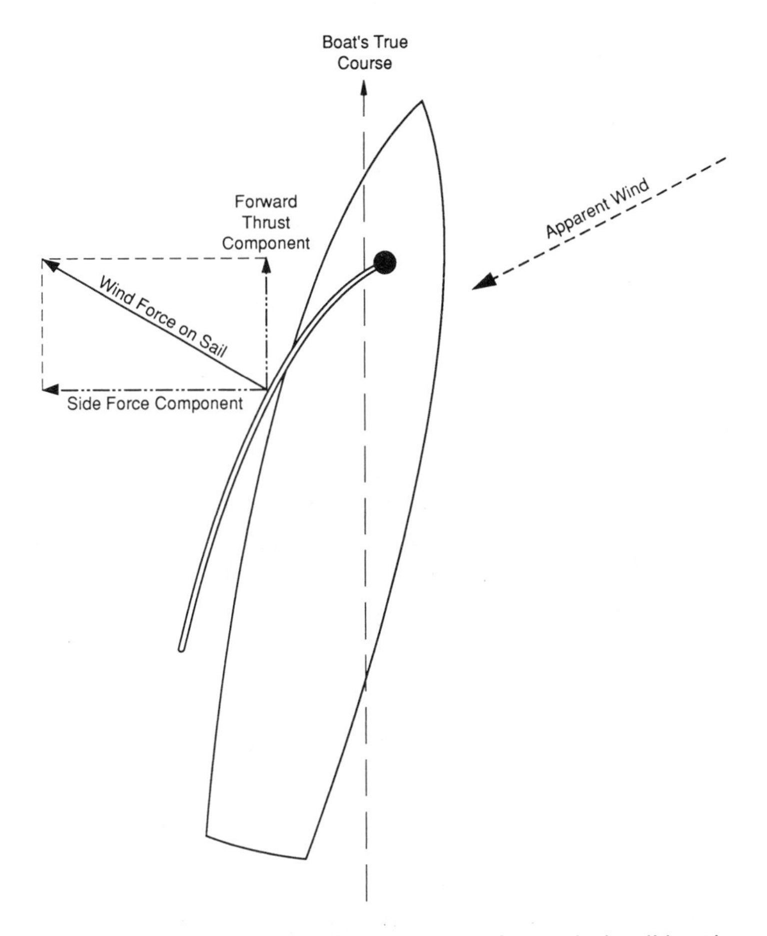

Fig. 11.2 Components of force arising from wind striking the sail (or sails).

2.2 Leeway. Figure 11.1 illustrates another important point. Notice that the boat's true course is somewhat downwind of the apparent course (as sighted along the boat's centerline). We speak of this as "making leeway," an action that results from the side thrust of the wind in the sails. The leeway angle (see figure) is usually rather small, typically less than 6 degrees.

2.3 Aerodynamics. How can a wind coming from over the bow allow a boat to move upwind? As air flows past a properly trimmed sail, pressures develop on both sides of the sail: positive pressure on the windward side and negative pressure on the leeward (pronounced "LOOerd") side. The net effect of these forces produces "lift," just as in an airplane wing. Figure 11.2 pertains to the net aerodynamic forces at play and is intended to clarify the concept. The diagonal arrow labeled *Wind Force on Sail* is a single substitute for the infinite number of individual forces pushing or pulling over the entire area of the sail. Its length and direction correspond to the cumulative total of all those forces. Part of that diagonal force tends to push the boat to leeward and that is labeled *Side Force Component* (also called *heeling force component*). The useful part of the diagonal force, shown as *Forward Thrust Component* (also called *driving force component*), urges the boat ahead.

Obviously, the boat cannot sail on a course directly up wind, but must be satisfied to steer at some angle off to one side or the other. It can reach an upwind destination only by following a zigzag course, a maneuver known as *tacking.* The most carefully designed racing boats can sail as close as about 35 degrees off the wind under ideal conditions of wind and waves. A typical cruising boat might hope to sail within 45 degrees of the wind. In heavy weather with large waves beating against the bow, the boat may experience so much side drift that the true course will be more like 90 degrees off the wind. In short, the boat will barely hold its own against the wind.

2.4 Hydrodynamics. Figure 11.3 introduces another set of complications: the hydrodynamic forces that counterbalance the pre-

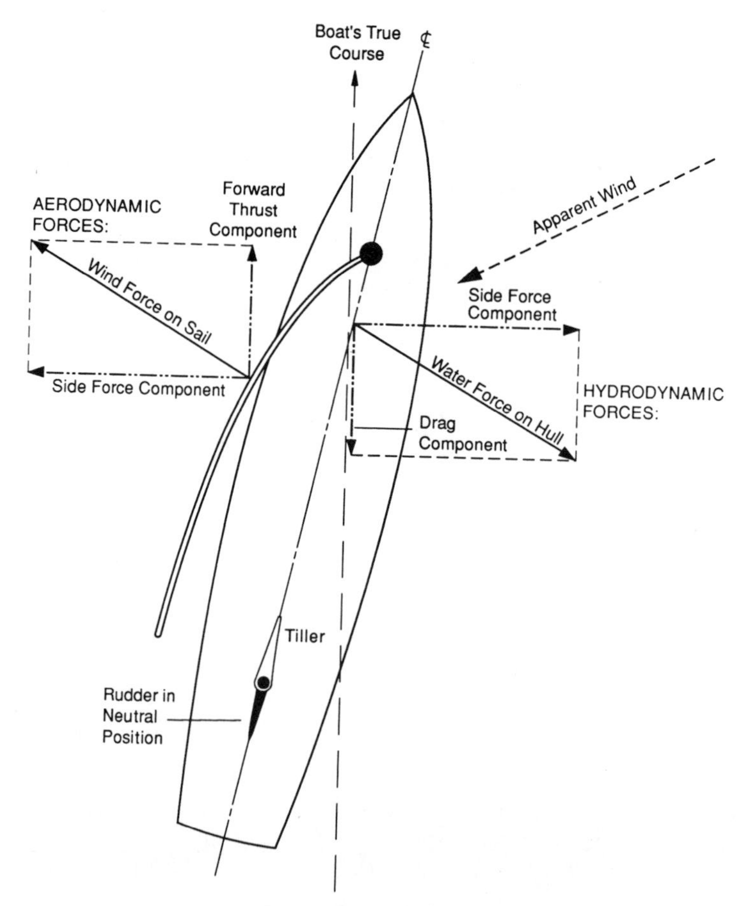

Fig. 11.3 Wind and water forces are exactly equal and opposite, but, being out of line, tend to turn bow into the wind.

viously discussed aerodynamic forces. Again, we substitute a single imaginary force, *Water Force on Hull,* that produces exactly the same total effect as the infinite number of individual forces arising from the action of the hull slipping through the water. This single force, as before, can be broken down into two components. The useful part, labeled the *Side Force Component,* resists the heeling force of the wind and effects a balance of lateral forces. This force component is produced by the inherent resistance of a hull to sideways motion, usually abetted by a keel or centerboard. The angle between the boat's centerline and the true course also contributes to the side force component. Indeed, without some angle of attack there can be no hydrodynamic side forces. As you may infer from the figure, the water impinging on the lee bow provides a side thrust upwind.

Unfortunately, there will always be a second, undesirable effect. In this case it is the *Drag Component* in opposition to the wind's driving force, shown as *Forward Thrust Component* in Fig. 11.3. If the wind picks up speed, the forward thrust on the sail will increase. That thrust will cause the boat to speed up until the new drag component equals the new forward thrust component.

Figure 11.3, being a bird's eye view, fails to show the rotational effect of the vertical spread between the side force components on sail and hull. That heeling moment is balanced by the righting moment generated by the heeled-over hull (as explained in Chapter VI), usually abetted by crew members climbing to the high side. Similarly, the vertical spread between the forward thrust component (on the sail) and drag component (on the hull) tends to trim the boat by the bow. In extreme cases, this trimming moment may cause the boat to bury its bow and perform a somersault. This catastrophic event is known as *pitch-poling.*

2.5 Balance. You will notice in Fig. 11.3 that the wind force and water force arrows are not in line with one another. The system as shown would therefore not be in balance but would tend to swing the boat's bow into the wind. The rudder, as shown in the drawing, is in the neutral position but can be turned to bring about the required balance of forces. Figure 11.4 shows the rudder turned so as to shift the water forces toward the stern until they are in exact alignment with the wind forces. To turn the rudder as shown, the person at the helm must push or pull a lever (called the *tiller*) upwind. If, as shown, the tiller must be turned "upwind" to effect a balance, we term that a *weather helm.* The opposite is a *lee helm.* Either way, the turned rudder will add drag and slow the boat. The aim in design, then, is literally and figuratively to steer a middle course and provide a *neutral helm* in average winds. In stronger winds the boat will heel more; the wind force will then be applied farther outboard and some degree of weather helm will be required. Conversely, in light airs some lee helm will be required.

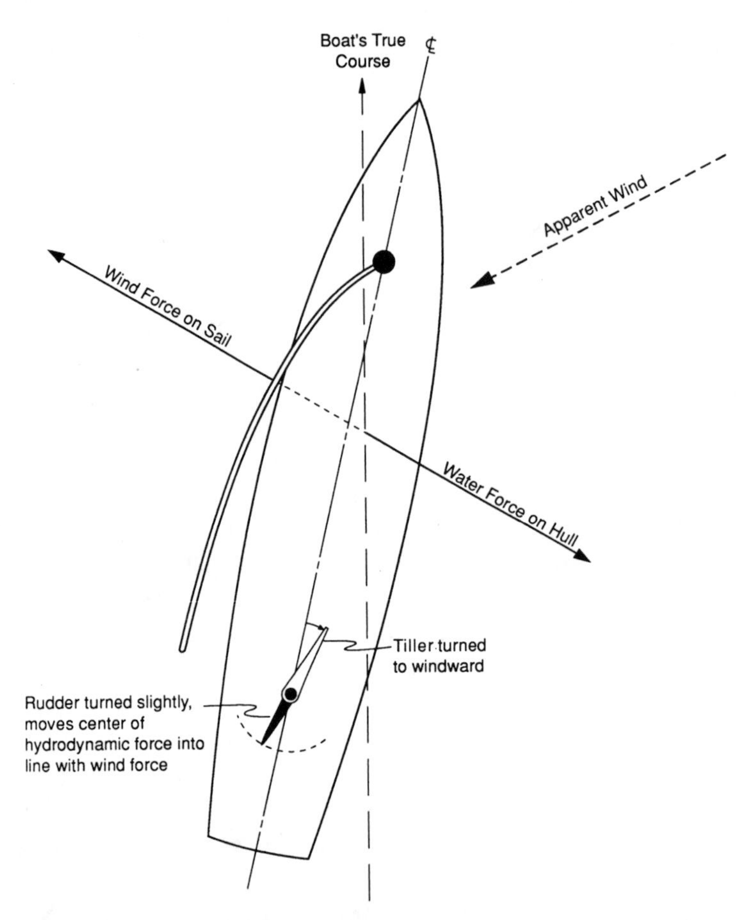

Fig. 11.4 The rudder may be used to bring wind and water forces into exact alignment and so keep the boat on a steady course.

In designing a sailboat, one of the naval architect's most ticklish jobs is to provide the neutral helm described above. Attaining a reasonable balance between wind and water forces is still mostly art, and occasional mistakes must be corrected by expensive changes in the finished boat. On some occasions, masts must be shifted or the keel profile may be changed. Because water is 800 times as dense as air, small changes in underwater profile are as effective as large changes in sail arrangements. On the other hand, modifying sail plans is usually much less expensive than changing the hull, so that is usually the place to start. If that doesn't do it, perhaps the mast should be raked or re-stepped. If the boat is still out of balance, then underwater alterations will be appropriate.

Perhaps you have mastered the skill of sailboarding. If so, all these questions of lateral balance are matters you have learned to handle instinctively. If you want to head the board more upwind, simply pull the mast and sail aft. To turn more downwind, push them toward the bow. The board's center of lateral resistance will remain the same but the center of lateral effort will change—and without benefit of rudder action.

3. Hull Forms

3.1 Monohulls. Sailboats should have hull forms that create low drag when sailing over a wide range of heel angles. This attribute is attained by adopting a generously curved transverse shape at mid-length, that is, a low maximum section coefficient. The prismatic coefficient will also be low and may range from 0.52 to 0.63, with median values favoring the lower figure. In short, sailboat hulls tend to be somewhat apple-shaped. Transverse stability is an important factor in the ability to carry sail. It can be attained either through adding ballast low in a narrow hull, or eschewing ballast, in whole or in part, in favor of wider beam. A wide beam tends to increase resistance, but that is offset by reduced displacement. Beamy boats offer extra deck area; they allow greater versatility in trimming sails and in arranging for efficient stays to strengthen the mast or masts.

Racing-boat hull forms are greatly influenced by rules of the various racing associations. To the extent that such influences decrease the boat's seaworthiness or comfort, the rules are to be deplored. Fortunately, in recent years rule reforms have made these pernicious influences less pronounced. The inherent problem is that the experts who invent the rules don't intend to encourage unseaworthiness, but they cannot foresee how future designers will torture the rules in their attempts to produce a winning boat. Moreover, experts cannot always agree on what makes one boat more seaworthy than another.

3.2 Multihulls. Catamarans are twin-hulled boats featuring widely spaced slim hulls with overall beam perhaps just under half the overall length. Trimarans have a moderately narrow center hull stabilized by outrigger hulls port and starboard. The typical trimaran has an overall beam somewhat over half its overall length. Both types of boats are characterized by large degrees of initial transverse stability in comparison to their displacement. As a result, multihull boats can carry a considerably greater area of sail than is possible with a monohull of equal weight. It follows that multihull boats are capable of speeds far in excess of those attainable by normal hulls. There are, however, disadvantages. Multihull boats usually have trouble completing the tacking maneuver when beating upwind. That is, owing to their light weight and awkward configuration, they are apt to swing into the wind, but then stall, pointing dead upwind like a weather vane and more or less out of control (a condition known as "being caught in stays"). Another major disadvantage is that their high initial stability is offset by their small range of stability. That is, if forced over very far by strong winds, a multihull boat is liable to flip over 180 degrees and then prove most awkward to turn back to the upright position. That is why a seagoing multihull boat may feature an escape hatch in the bottom. A third drawback is the matter of structural complications that become apparent when a multihull boat encounters waves. Large twisting forces may come into play and structural failure may ensue. A final negative factor is the danger of *pitch poling,* explained in Section 2.4 above.

4. *Typical Sail Arrangements*

4.1 Standard rigs. Figure 11.5 shows simple sketches of six common sailboat rigs. The four-sided sails have a spar called the *gaff* at their tops and are therefore called *gaff sails.* The triangular sails are variously referred to as *jib-headed, Marconi,* or *Bermuda* sails. These two kinds of sails can, in general, be used interchangeably. That is, a ketch can as well have one kind of sail as the other. The yawl and the ketch are differentiated by the relative size and location of the small sail at the stern. This is called the *mizzen sail.* The mizzen sail on a yawl is relatively small compared to that on a ketch and it is located farther aft, usually aft of the rudder stock. The ketch's relatively larger mizzen sail is carried on a mast that is located forward of the rudder stock.

4.2 Rigging terminology. Sails are raised by a rope called the *halyard* or *halliard.* (These terms date back to the days of square-rigged

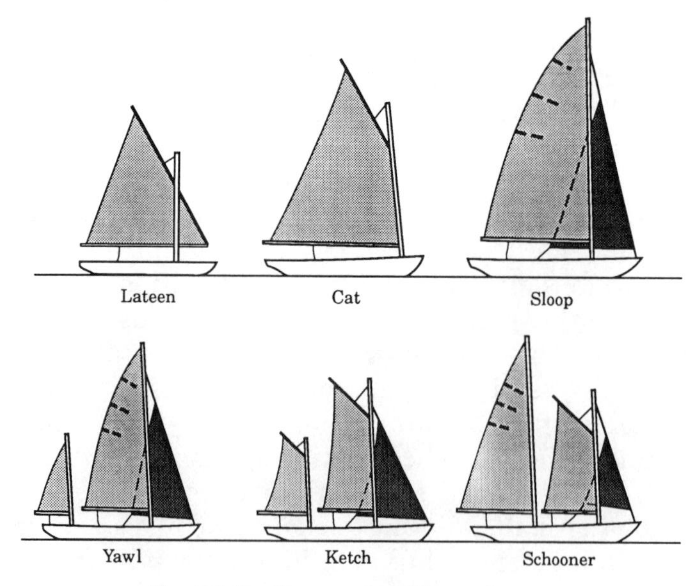

Fig. 11.5 Common sailboat rigs.

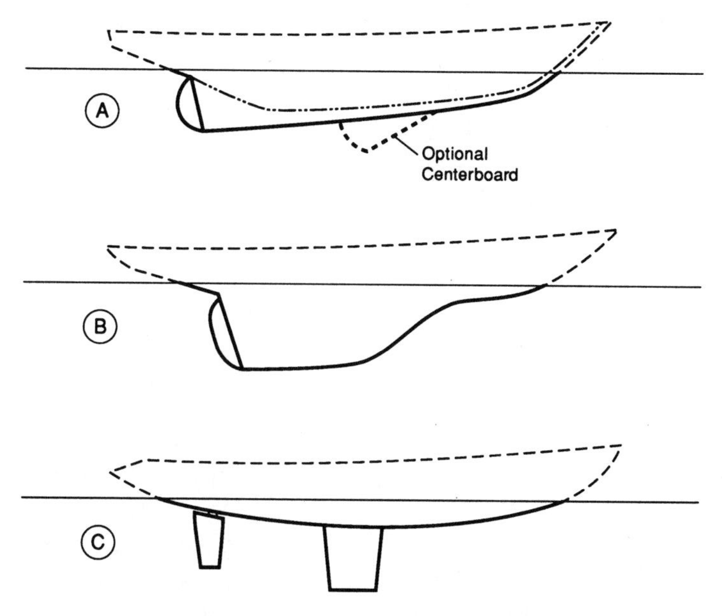

Fig. 11.6 Typical sailboat underwater profiles.

ships when the sails were hoisted by hauling up on the yards to which they were attached.) The lines that are used to control the position of the sail are called *sheets,* much to the confusion of every beginner. The spar at the bottom of a sail is the *boom.* Portable spars used for various purposes are held in place by *guys.*

The *standing rigging* comprises those parts of the sailing gear that are more or less permanently fixed in position. Included here would be the masts and their supporting wires or ropes. *Shrouds* provide lateral support; *stays* provide fore-and-aft support. All else is *running rigging,* principally halyards, sheets, and guys, as well as the sails themselves.

In the case of sail boards, all elements of the rigging (and crew) are integrated into a well-coordinated unit. Standing rigging and running rigging are one, while crew and ballast are also fully integrated.

5. *Underwater Profiles*

Figure 11.6 shows three typical sailboat underwater profiles. Sketch A is representative of older yachts. The keel is little more than a shallow bar, rectangular in section, hung below the hull proper. It deepens at the stern in a region termed the *deadwood* so as to give support to the rudder. Resistance to leeway may be increased by the addition of a *centerboard,* a flat, hinged appendage that may be pulled back up into the hull for shallow water operations.

Sketch B is typical of modern racing or cruising sailboats. The deep keel probably carries considerable weight of fixed ballast. Such a boat is appreciably better at working its way upwind than the boat shown in A, and its deeper, narrower rudder can provide equal turning force with less wetted surface, hence less frictional resistance. This hull, too, may be fitted with a centerboard (not shown).

Sketch C shows a profile appropriate to a light-displacement boat with wide beam but little or no ballast. The keel is so narrow fore and aft that the rudder must now be hung closer to the stern as a separate appendage. If the rudder were to be fitted to the keel, it would be too close to the boat's turning center to provide much of a turning moment.

A *wing keel* has inverted V-fins (called *winglets*) fitted to its bottom. Such added appendages are intended to slow water from flowing under the keel. As such, they are supposed to enhance the boat's hydrodynamic resistance to leeway. They are used to best advantage where draft limitations force keels to be relatively shallow. An important secondary advantage may lie in their ability to increase transverse stability. If made of lead, their low position will lower the boat's center of gravity. Where deeper keels are possible, winglets may be disadvantageous because they add somewhat to drag.

6. *Commercial Sail*

6.1 Pertinent technological developments. With the advent of suddenly raised oil prices a couple of decades ago came a renewal of interest in a return to wind propulsion for merchant ships. The wind is a free and inexhaustible source of nonpolluting energy that is also free of political manipulation or control. Moreover, since the time a century ago when steam was replacing sail, we have seen the development of several technologies that would strengthen the case of the modern wind-powered ship. Among these technologies are aerodynamic principles applied to sail design, new lightweight materials for masts and sails, weather routing, and satellite communication systems.

As an incidental note here, I think it worth mentioning that in the third world commercial sailing ships are still much in evidence. There are today, for example, something like 4000 modest-sized, wooden-hulled sailing ships serving the inter-island commerce of Indonesia.

6.2 Benefits of size. There is a theoretical scale factor in favor of large size in sailing ships. I refer to the fact that the heeling moment on a sailing vessel varies as the sail area times the heeling lever, that is in proportion to the scale cubed. The hull's righting moment, on the other hand, varies as the displacement times the righting lever (remember *GZ* in Chapter VI?). This means that the righting moment varies with the scale raised to the fourth power. To cite an example, if you have one sailboat of a given design and another that looks exactly the same but is twice as big in every dimension, the scale factor would be two. In a wind of a given speed, the larger vessel's heeling moment would exceed that of the smaller vessel by a factor of two cubed, or eight. The righting moment, on the other hand, would exceed that of the smaller vessel by a factor of two raised to the fourth power, or sixteen. In short, the larger vessel can safely carry proportionately far larger sails. This concept, and the matter of wave-making resistance, explain why big sailboats are almost always far superior in speed to smaller boats.

6.3 Limits to size. Unfortunately, sailing ships large enough to be economical cargo carriers possess inherent disadvantages. For instance, the standing rigging becomes enormously complicated, heavy, and expensive. Then, of course, there is the matter of bridges that must be sailed under to enter most harbors. The scale advantage explained in the paragraph above is negated by these practical technicalities. This leads us to believe that commercial sail can best be applied to ships of relatively modest size. (Don't look for sails on supertankers.)

6.4 Best applications. Another obvious disadvantage of wind power is its unreliability. Ships intended for liner operations (that is, sailing on a fixed schedule) would therefore be ill-suited for sail propulsion. One can conclude that the ideal application would be for relatively small bulk carriers on long ocean voyages. Bulk carriers seldom operate on a fixed schedule, and on long voyages the saving in necessary fuel weight can be converted into added cargo capacity. There remains the problem of finding a rigging configuration that will not unduly inhibit the cargo-handling operations at either end of the voyage, but one may expect that satisfactory solutions could be found for that problem.

6.5 Auxiliary power. Most proponents of commercial sail recognize that there are times when some auxiliary, mechanical propulsion system is desirable. Entering or leaving port and operating near the equator (where winds are less than hearty) make such mechanical power an economic necessity. In short, no one is advocating a return to pure sail. Some small diesel engine is a basic requirement.

6.6 A secondary role for sail. Given all of the above considerations, there seems to be little cause for expecting any near-term revival of commercial sailing ships serving the commerce of developed nations. There is, however, some reason to consider the use of sails as auxiliary power for ships that rely primarily on mechanical propulsion. If the wind is there, why not raise a sail or two and let it help either to increase speed or save fuel? Figure 11.7 shows how ships with lengthy deck cranes can rather easily hoist sails to exploit a favorable wind.

Figure 11.8 shows a more sophisticated form of auxiliary sail on a

Fig. 11.7 Crane-equipped cargo ship adapted to carry auxiliary sails.

ship with specially fitted masts and sails. The entire rig is arranged for mechanical operation and remote control. The masts are rotated by motors so as to trim the sails to suit the apparent wind direction. The fabric sails are supported by metal frames that have vertical hinges. These allow the sails to be folded back around the masts when not in use. This kind of rig is more expensive than that shown in Fig. 11.7; it does, however, avoid the disadvantage of demanding extra work from the crew.

Auxiliary sails are also sometimes found on cruise ships. Many people like the idea of taking passage on a sailing vessel. The sails not only save fuel but also tend to reduce rolling and vibrations and thus offer a more comfortable ship. Figure 11.9 shows one form of auxiliary sails fitted on a large cruise ship.

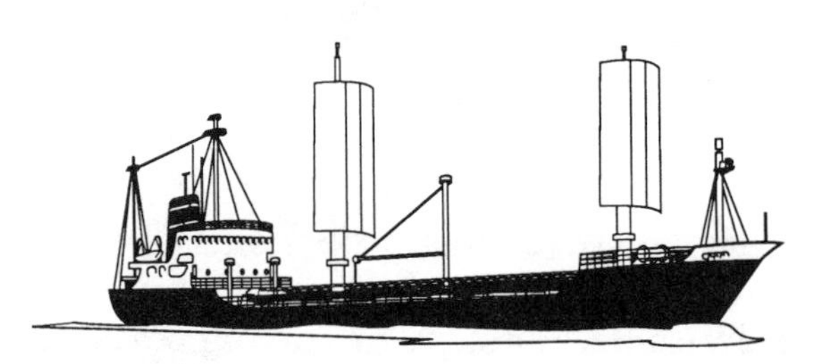

Fig. 11.8 Mechanically operated auxiliary sails fitted to a small tanker.

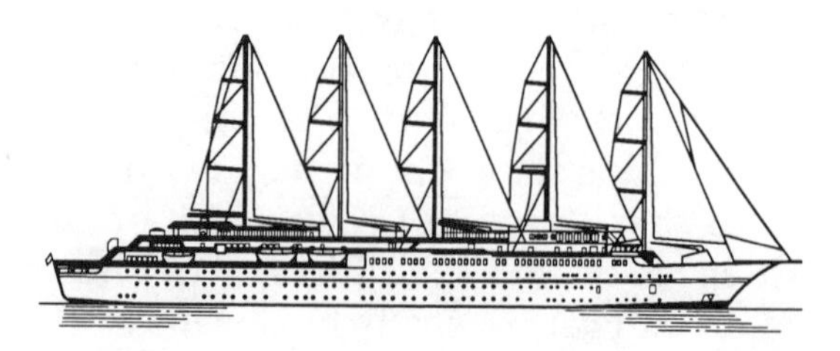

Fig. 11.9 One form of auxiliary sail applied to a cruise ship.

Further Reading

Henry, Robert G. and Miller, Richards T., "Sailing Yacht Design—An Appreciation of a Fine Art," *Transactions,* Society of Naval Architects and Marine Engineers, Jersey City, N.J., 1963, pp. 425–490.

Letcher, John S., Jr. et al., "Stars & Stripes," *Scientific American,* Aug. 1987, pp. 34–40.

Kinney, Francis S., *Skene's Elements of Yacht Design,* Dodd, Mead & Company, New York, 1973.

Hamlin, Cyrus, *Preliminary Design of Boats and Ships,* Cornell Maritime Press, Centreville, Md., 1989.

Marchaj, C. A., *Aero-Hydrodynamics of Sailing,* Dodd, Mead & Company, New York, 1980.

Herreshoff, Halsey C., "Hydrodynamics and Aerodynamics of the Sailing Yacht," *Transactions,* Society of Naval Architects and Marine Engineers, Jersey City, N.J., 1964, pp. 445–492.

Kirkman, Karl L. and Pedrick, David R., "Scale Effects in Sailing Yacht Hydrodynamic Testing," *Transactions,* Society of Naval Architects and Marine Engineers, Jersey City, N.J., 1974, pp. 77–125.

Bobrow, Jill and Jinkins, Dana, *The World's Most Extraordinary Yachts,* W. W. Norton, New York, 1986.

Mate, Ferenc, *The World's Best Sailboats, A Survey,* W. W. Norton, New York, 1986.

Daniels, Jane, *The Illustrated Dictionary of Sailing,* Gallery Books, New York, 1989.

Miller, Richards T. and Kirkman, Karl L., "Sailing Yacht Design—A New Appreciation of a Fine Art," *Transactions,* Society of Naval Architects and Marine Engineers, Jersey City, N.J., 1990, pp. 7-1 to 7-39.

Fig. 11.10 Sailing craft extremes. Upper: U.S. Coast Guard's square-rigged training bark *Eagle.* Courtesy R.W. Paugh, U.S. Coast Guard; photo by Robert Basten. Lower: Lester Rosenblatt's ketch *Rosa II* in a fresh breeze. Courtesy M. Rosenblatt and Son, Inc.

CHAPTER XII

HULL STRENGTH REQUIREMENTS

1. *Introduction*

Build me straight, O worthy Master!
  Stanch and strong, a goodly vessel,
That shall laugh at all disaster,
  And with wave and whirlwind wrestle!

Those lines, which introduce Longfellow's poem "The Building of the Ship," make a fitting introduction to this chapter. (Perhaps I should first explain that Longfellow used the word *stanch* in the archaic sense of meaning both reliable and proof against leakage.) In this chapter I explain what kinds of loads a vessel may encounter in service. I also explain how naval architects go about assuring a hull that will be strong enough to resist those loads—but still be not overly-strong (which means overly-heavy and expensive).

In Chapter XIII you will learn something about the major kinds of shipbuilding materials, and then Chapter XIV goes into the subject of structural components and how they are arranged to guarantee that the vessel can "with wave and whirlwind wrestle," and win every time.

2. *Functions*

2.1 Resisting local loads. Wherever you look on a boat or ship you will see local loads. When you step on the deck your weight is imposing a load on it. And that load is carried through the hull structure right down to the bottom, where the buoyant forces of the water provide equal and opposite support (itself a form of load). The mooring lines are exerting loads on the mooring fittings. Mast stays are pulling up on the sides of the boat, while the mast is thrusting down on the bottom. Figure 12.1 shows cargo pushing down on a deck, while water pressure under the hull opposes the weight of the cargo and imposes loading on the bottom structure. The water pressures on the sides of the vessel are equal and opposite, but must be withstood by the hull structure. These are but a few examples of what we mean by local loads.

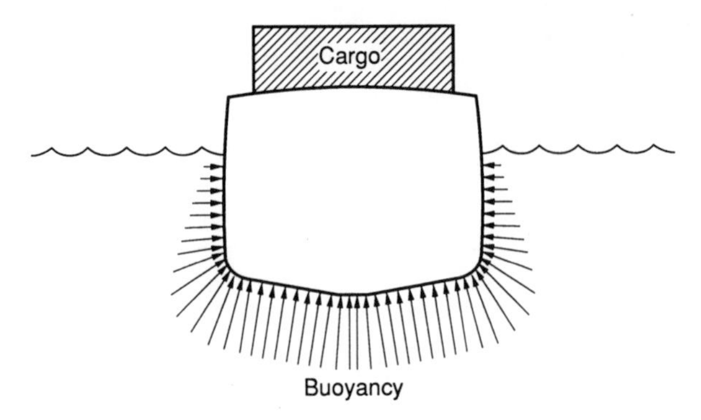

Fig. 12.1 Cargo weight supported by buoyancy.

2.2 Resisting loads spread over the ship's length. In addition to withstanding local loads, ships must be given enough overall strength to prevent their breaking in two. This is not terribly likely in small craft, but is of major concern in ships of any appreciable size. Naval architects think of the hull as a big box girder deriving its major strength from the bottom shell, side shell, and uppermost continuous deck. Predicting as best they can the overall loads that may be imposed in service, naval architects specify plating of sufficient strength to withstand those loads with modest margins for safety and for a lifetime of gradual corrosion. Safety margins are necessary because predicting the maximum load that may be imposed on a ship throughout its many years of life is hardly an exact science.

Figure 12.2 shows the profile of a simple rectangular barge subjected to five different longitudinal loading conditions with cargo car-

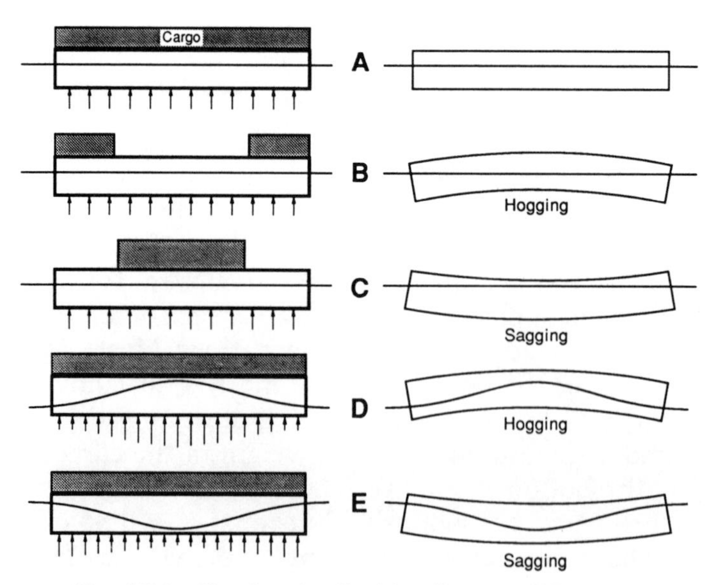

Fig. 12.2 Five longitudinal loading conditions.

ried only on the upper deck. In the first three sketches the barge is floating in still water. In Sketch A the deck cargo is spread uniformly over the length of the barge. That being the case, the weights and supports (that is, buoyancy) are all perfectly aligned. There would be local loads, but there would be no tendency for the barge to deflect from end to end, or to break in two.

In Sketch B the cargo is divided in two, with half moved to each end of the barge. Now the weights and supports are out of line and the ends of the vessel will tend to droop, a condition known as *hogging*. The extent of the hogging is exaggerated in the sketch on the right, but the possibility of breaking in two is real, nevertheless.

In Sketch C the cargo is shown concentrated within the barge's midship half-length. The weights and supports are again out of line, but now the deflections will be concave upward, a condition called, appropriately, *sagging*. Once more, the extent of the distortion is exaggerated, but complete hull failure is a distinct possibility.

In the first three sketches we assumed the barge was floating in still water, and you could visualize it breaking in two simply because of unwise cargo distribution. As you may readily imagine, when the barge encounters waves, the buoyant support will no longer be uniform, so the misalignment of weights and supports may become worse. In Sketch D the deck cargo is again spread uniformly over the length of the barge, but now we have a wave with its crest at mid-length and hollows at the ends. There is excess buoyancy at mid-length but deficient at the ends. Under those conditions hogging will occur. If the loading condition of Sketch B were to coincide with the wave condition of Sketch D, an even worse hogging condition would result.

Sketch E shows a wave with crest at each end of the barge and hollow at mid-length. A condition of sagging occurs, and that would be exacerbated if the loading condition of Sketch C were superimposed.

From all this you may see that overall longitudinal loads arise because of misdistribution of weights and supports over the vessel's length. Before explaining the techniques for designing against complete hull failure in the face of such loads, I want to give you a brief introduction to some of the theoretical bases of the analysis.

3. *Modes of Failure*

3.1 Bending. The term *beam* has several meanings. In the most general sense, a beam is any structural member intended to carry loads that are out of alignment with the supports. Within a hull, many specific structural members act as beams. These include deck beams, shell frames, bulkhead stiffeners, and various secondary supporting members referred to as girders, webs, and so forth, depending on their location. Sketch A in Fig. 12.3 shows a sample case: a beam with a single load at mid-length and so-called simple supports at each end. (*Simple supports* are of such a nature that the ends of the beam are free to tilt, as shown in the sketches.)

Under most conditions of loading, the worst stresses in a beam are imposed at right angles to the direction of loading. As a result, failure may occur as shown in Sketch B. The bottom of the beam is pulling apart because of excessive tension, while the top crushes together because of excessive compression. This beam has failed because of excessive *bending moments.* The loads and supports being so far apart have bent the beam much as a pair of hands (Sketch C) might bend, and possibly break, a metal bar. Bending moments have two components because they involve forces acting at a distance. Their usual British units, then, are foot-pounds.

3.2 Shear. Failure may also occur because of what we call *shear.* Sketches D and E show two kinds of shear failure: "normal" shear (Sketch D) and longitudinal shear (Sketch E). I don't intend to say much more about shear failure because bending moments (Sketches B and C) rightfully demand the most attention in ship design.

3.3 Stress and strain. Before going on, I want to explain the difference between two technical terms that are often confused in

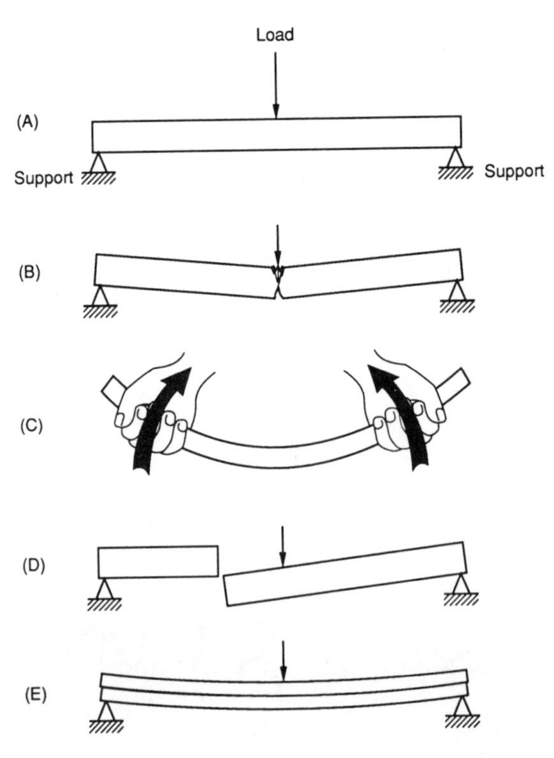

Fig. 12.3 A beam with simple supports.

everyday conversation. *Stress* measures forces per area of cross section, whether in tension, compression, or shear. In British units these would be in pounds per square inch (psi). *Strain* is a nondimensional measure of elongation, or contraction, found by dividing the component's stressed length by its original, unstressed length.

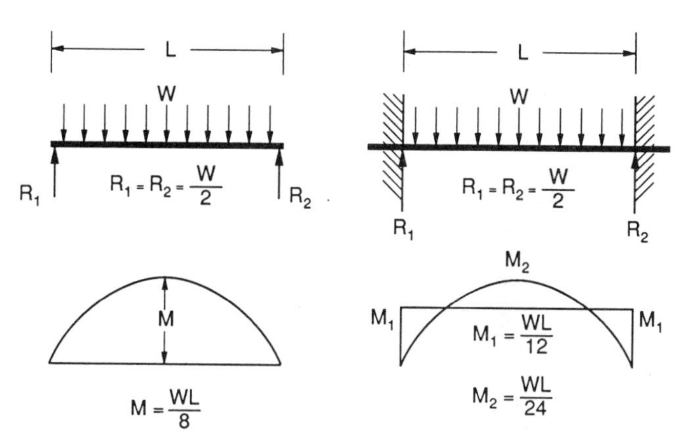

Fig. 12.4 Beam diagrams.

4. *Beam Loading*

4.1 Beam diagrams. In order to predict the stresses in a beam, the naval architect must first predict how great the loads will be and where they will occur. This information leads to a determination of what the resulting bending moments will be. Depending on the structural component under study, this prediction may be fairly simple or extremely complex. Figure 12.4 shows two simple cases of uniformly loaded beams with supports at each end. The beam on the left has simple (free-end) supports; that on the right has built-in supports, more or less as might occur if brackets locked the beam into some rigid adjacent structure. In real life, neither condition occurs pure and simple. Even without brackets the ends of most members are welded to some other structure and so are not entirely free to pivot. Where brackets are fitted they are never attached to perfectly rigid structures because on ships or boats there is none such.

In Fig. 12.4 W stands for the total weight on the beam (to simplify the analysis, we usually assume the beam itself is weightless). R stands for the reactions (that is, resulting loads) on the supports. Because of symmetry, the reactions at each end are the same, and this is true for both free-ended and bracketed beams.

A straightforward mathematical analysis, which we shall not go into here, leads to the equations for bending moment values shown in the figure.

4.2 Benefit of brackets. The curves shown in the lower part of Fig. 12.4 indicate the bending moment at each point along the length of the beam. In the simple-supported beam, there can be no bending moments at the ends because they are free to pivot. The maximum moment occurs at mid-length, and that is where the beam would be most likely to fail. In the bracketed beam, the maximum bending moments occur at the extreme ends. If you compare the two curves, you will see an important difference, which is that the brackets reduce the maximum bending moment to two-thirds of that imposed on the beam with simple supports: ($WL/12$ instead of $WL/8$). This means that, for a given load, the bracketed beam needs to be only two-thirds as strong as the other. Fitting brackets, then, can lead to important savings in the weight and cost of hull framing members, which usually more than offset the added labor required to fabricate, fit, and weld the brackets.

5. *The Hull as a Girder*

5.1 Classical modes of failure. In Fig. 12.5, Sketch A shows a ship that has encountered a wave of its own length and with a crest at mid-length. This has caused a failure in the hogging condition. The deck has failed in tension and the bottom structure has failed by buckling in compression. The line NA is the boundary between where the structure is in tension and in compression. We call this boundary the *neutral axis.*

Sketch B shows the opposite kind of classical failure. Here, wave crests at the ends of the ship have caused a failure in the sagging condition. In this case, the deck has failed in compressive buckling

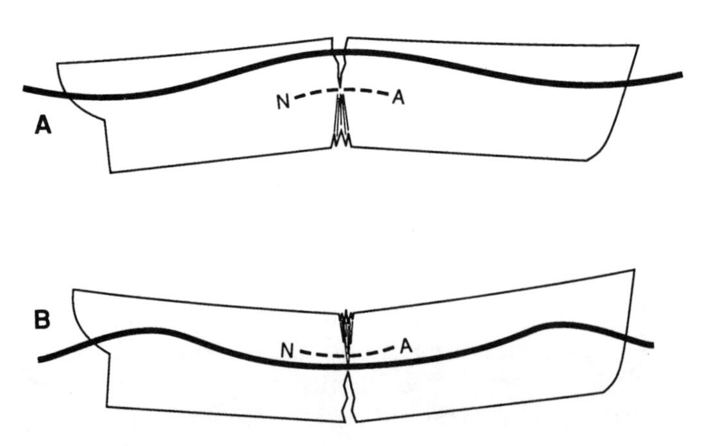

Fig. 12.5 Classical modes of hull failure owing to excessive longitudinal bending moments.

and the bottom has pulled apart in tension. Once again there is a neutral axis where the hull is subject to neither tension nor compression. The location is exactly the same whether the vessel is hogging or sagging. Section 6 explains how to find the neutral axis.

5.2 Brittle fracture. During the Second World War, when the first all-welded ships were built, a new and unexpected mode of failure made itself known. Because of flaws in design or welding, modest levels of tensile loading could start a crack in the top or bottom structure. Such a crack might then almost instantly continue right on around the hull. This resulted in an all-tensile failure, with no buckling at all. Such failures seldom occurred in riveted ships because a crack, upon reaching the edge of one plate, would spread its stresses through several rivets in the seam joining that plate to the next. The riveted seam, in short, acted something like a shock absorber. In welded ships, on the other hand, when a crack reached a welded seam there was nothing to prevent its continuing right on through to the next plate and then all subsequent plates.

Another factor was the manner in which the individual structural members pulled apart. Normally, if a member is overstressed in tension its ductility will cause it to "neck down" before it pulls apart. (Picture yourself pulling taffy.) This ductile action may relieve the stress in an overloaded component and allow it to pass some of its load on to its neighbors. In the all-welded ships that failed, however, investigators noted no evidence of necking down. Once the fracture started, it continued so rapidly that there was no time for ductile distortion. The steel plates and shapes had behaved like glass and had suffered *brittle fracture.*

The problem of brittle fracture is now essentially a thing of the past. Tougher steels are specified and more care is applied in design and workmanship so as to avoid stress-raising features.

6. The Neutral Axis

6.1 Fundamental concept. Theoretical analysis shows that the neutral axis occurs exactly at the center of area (or gravity) of the cross section of the beam. That is, when loaded in bending, all the material on one side of the center of gravity will be in tension; all that on the other will be in compression. In Fig. 12.6, Sketch A shows an I-beam. Owing to symmetry, the center of gravity, and hence neutral axis, is exactly at mid-depth. In an ordinary angle, Sketch B, the neutral axis is below mid-depth. In a bulb angle, Sketch C, the neutral axis is higher than that in Sketch B because adding the bulb has shifted the center of gravity.

6.2 Plates and shapes in combination. In Fig. 12.6, Sketch D shows an angle riveted to a plate. With the plate and angle acting

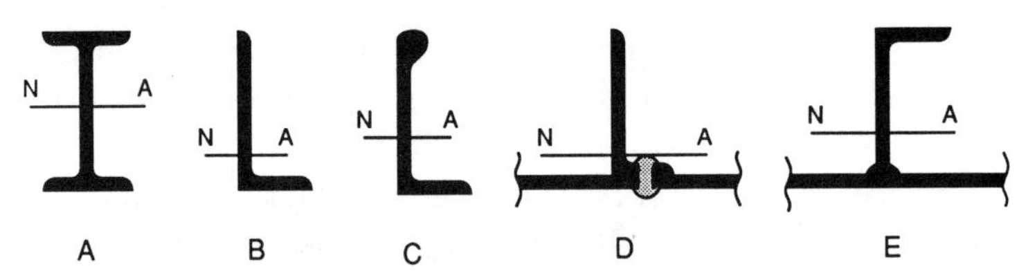

Fig. 12.6 Locating the neutral axis.

together as a beam, the neutral axis is close to the plate. As will be shown later, material close to the neutral axis contributes little to the strength of a beam. This means that, while the short, riveted leg of the angle is necessary for holding rivets, it contributes little to the strength of angle-and-plate as a beam. Sketch E shows the modern, welded equivalent of Sketch D. Here the angle is inverted and welded to the plate. That shifts the center of gravity, increasing the effectiveness of both the plate and the short leg of the angle. This illustrates one of the advantages of welding over riveting.

I have been describing all of the above in terms of vertically loaded beams. The same logic applies regardless of the direction of loading.

6.3 Neutral axis of the hull. To locate the neutral axis of the hull, naval architects once again look for the center of gravity, but now they must be selective. The center of gravity of exactly what? They concentrate their attention only on those parts of the hull structure that contribute directly to the bending strength of the hull girder. What these include are members that run continuously fore and aft throughout at least the ship's midship half-length (which is pictured in Fig. 12.9). The analyst specifically omits all framing members that run transversely and all longitudinal members that are discontinuous (for example deck structure forward or aft of hatch openings).

The mechanics of this analysis are simple in principle. The naval architect makes a table which records, for each part of the contributing structure, its cross-sectional area and distance from its center to the ship's baseline. The analyst calculates the product of each area times its distance, giving its moment about the baseline. All areas and all moments are summed. The total moment divided by the total area gives the vertical center of gravity of the contributing members. That, in turn, gives the vertical location of the neutral axis. As will be shown in the following sections, this is an essential step in assessing the ability of the hull girder to withstand the loads imposed upon it and so resist breaking in two. Today these analyses are usually performed by computer.

7. *Stiffness*

7.1 Moment of inertia. The stiffness of a beam is directly influenced by a property of its cross-sectional geometry that we call its *moment of inertia.* (A more descriptive term sometimes used is *moment of resistance.*) How far a beam will deflect under any given load is directly proportional to its moment of inertia. This, of itself, is not too important. As we shall see, however, finding the moment of inertia is a necessary step in predicting bending stresses in the hull girder. The fundamental way to find a moment of inertia is to take each component of the cross section and multiply its area by the square of its distance to the neutral axis, and then add up all those products. Without going into the details of this subject, Fig. 12.7 gives a sense of how cross-sectional shape will affect moments of inertia. The moments of inertia of rolled shapes, such as the I-beam shown in the figure, are found in standard handbooks published by steel or aluminum companies. Also note that we use a capital I as our symbol for moment of inertia as well as for a beam of that cross-sectional shape.

7.2 Moment of inertia of the hull girder. Simply put, the naval architect who wants to compute the ship's moment of inertia must carry out a complex series of simple individual steps. Once again the analysis (usually done by computer) applies only to those parts of the structure that are involved in finding the neutral axis. Each individual area is multiplied by the square of its distance to the neutral axis. Moreover, additions are made for the moment of inertia of each vertical member (such as a side shell plate) about its own neutral axis.

7.3 Stiffness of the material. How far a beam will deflect also depends on its material of construction. The technical term for this property is the *modulus of elasticity,* sometimes called *Young's modulus.* (The term *modulus* means simply "measure.") You can think of the modulus of elasticity as being the stress that would double the length of a sample of the material *if it did not break in two* along the way. For steel alloys the value is close to 29 million pounds per square inch. For aluminum alloys, Young's modulus is about one-third that value, and for fiberglass, about one-tenth.

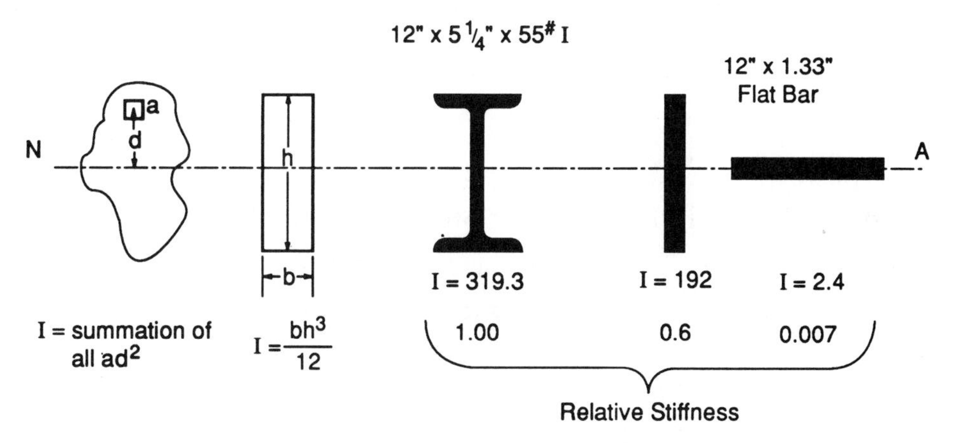

Fig. 12.7 Moments of inertia and relative stiffness.

8. *Section Modulus*

8.1 Basic concept. Stiffness is almost always a virtue in a beam, so a high value of the moment of inertia is good. That value, however, must be achieved without allowing the most highly stressed parts of the beam to become overstressed. And where are the most highly stressed parts? Returning to Fig. 12.5, you may see that those members that are farthest from the neutral axis are the ones that are deflected the most and thus experience the highest stresses. We refer to them as the *outermost fibers,* and the distance from the neutral axis to those outermost fibers is commonly symbolized by the letter y. Looking strictly at a beam's cross section, then, we can see that its ability to resist bending moments will be directly proportional to its moment of inertia, I, and inversely proportional to the distance from its neutral axis to the outermost fiber, y. From this we can conclude that the overall measure of a beam's ability to sustain a load is equal to the ratio of I to y. This ratio is called, appropriately, the *section modulus.* It is often symbolized with a capital S, but may also be left simply as the fraction I/y (spoken of as "I over y").

8.2 Finding section modulus. Numerical values of the section modulus of standard rolled sections can be found in handbooks published by steel or aluminum companies. For nonstandard or complex sections, simply compute the moment of inertia, as before, and then divide by the distance y. This latter concept, as you might guess, applies to the hull girder as well. The neutral axis of the hull girder is seldom at mid-depth. The naval architect must therefore consider two y-values: that to the deck and that to the bottom shell.

8.3 Deriving stress. The maximum bending stresses that occur in a beam (at the outermost fibers) are directly proportional to the maximum value of the applied bending moment and inversely proportional to the section modulus. Indeed, it can be shown that the maximum stress is exactly equal to the maximum bending moment divided by the section modulus.

Maximum stress = Maximum moment ÷ Section modulus

$$s = \frac{M}{I/y}$$

In structural analysis, this equation is paramount.

9. *Analyzing the Hull Girder*

9.1 The midship section. The longitudinal strength of the hull girder is based on a drawing called the *midship section.* This is an imaginary cross section of the hull, which, to play it safe, combines the two weakest features of the design. The depth of the hull is taken at its minimum location (usually at mid-length, where the sheer curve is at its lowest point), while the structural details are taken at their weakest point, usually at a hatch opening. This drawing, incidentally, is usually the first one the naval architect prepares in designing the hull structure.

9.2 Finding section modulus. To assess the overall strength of the hull girder, the naval architect applies the principles outlined above to find the section modulus of the structure shown in the midship section. Again, only those parts of the structure that contribute directly to the longitudinal strength are taken into consideration. The step-by-step procedure is to (a) find the neutral axis, as explained in Section 6, (b) solve for the moment of inertia, *I*, as explained in Section 7, and, finally, (c) find the section modulus by dividing the moment of inertia by the distance from the neutral axis to the outermost fiber, as explained in Section 8.

9.3 Predicting maximum bending moment. Now comes a most complicated multistep procedure, that of analyzing the worst-case combination of loads pushing down on the hull and buoyant supports pushing up. The first step is to draw a curve that shows the weight-per-foot of the fully laden ship at each point along its length. See Sketch A in Fig. 12.8. The second step in the traditional method is to create an imaginary wave of length equal to that of the ship, with a crest at each end and hollow at mid-length. (The "standard" wave that is assumed here is a trochoid with length-height ratio of about twenty to one, depending on ship's length.) From this, and usually employing a special computer program, the naval architect uses a trial-and-error procedure to give the ship the combination of draft and trim that will make the buoyancy equal the total weight and also place their longitudinal centers in alignment. The resulting buoyancy curve is then drawn on the same baseline as that used for the weight curve, as shown in Sketch B of Fig. 12.8. As you will note, the weight exceeds the buoyancy at the ship's mid-length, while the buoyancy exceeds the weight near the ends. This misdistribution of weights and supports leads to the bending moments that may break the ship in two. Sparing you further details, let me simply say that after four more steps (made less arduous with computer assistance) the naval architect can derive a curve of bending moments at every point along the length of the ship (as shown in Fig. 12.9). This leads to a reliable estimate of the worst sagging bending moment the ship is likely to experience in service. Next, the naval architect repeats the above procedure, but now the ship is poised with the crest of the wave at mid-length. This leads to a reliable estimate of the worst hogging bending moment.

The prudent naval architect may go through all of the above analyses over and over again, but each time with different assumptions as to the distribution of weights along the length of the ship.

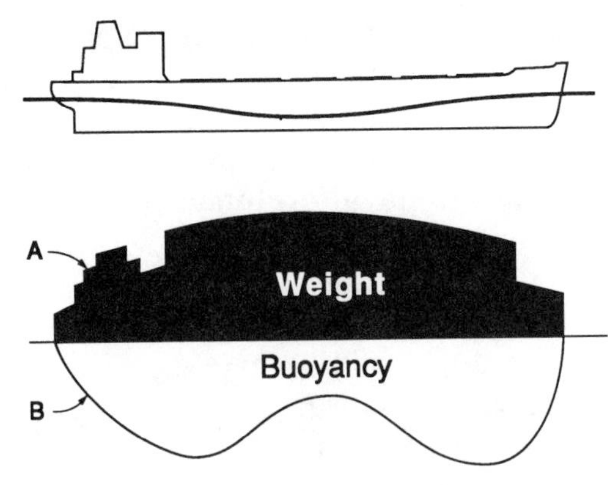

Fig. 12.8 First steps in predicting bending moments.

Are you curious about what is meant by a "trochoidal" wave? A *trochoid* is a curve that is traced by a point on the radius of a circle as the circle is rolled along a flat surface. Such a curve is closely akin to real ocean waves in that its crests are somewhat sharper than those of a true sinusoidal curve (which is what is traced by a point on the rim of the circle).

9.4 Predicting maximum stresses. To this point the naval architect has predicted the maximum bending moments that the ship may encounter both in hogging and in sagging conditions. The naval architect has also found the section modulus for the hull girder. Now all that remains is to divide the maximum moment by the section modulus, which leads to a reasonably reliable prediction of the maximum stress that the hull girder is likely to encounter during the life of the ship. Naval architects have enough humility to admit that they cannot

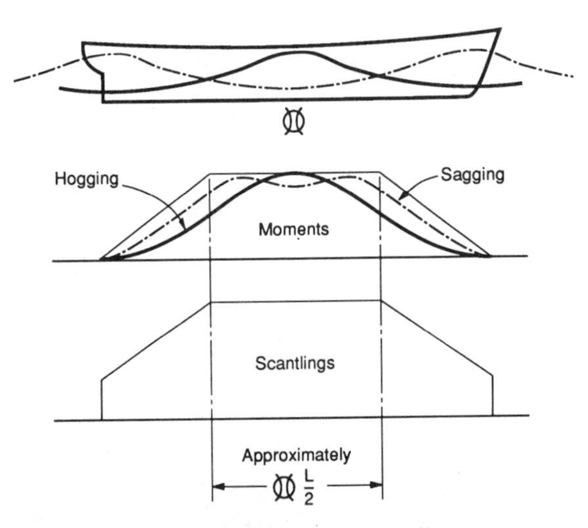

Fig. 12.9 How longitudinal bending moments influence the distribution of structural strength over the length of the ship.

know for sure what worst conditions of loading and waves will be imposed on a ship over its life. Thus, some prudent factor of safety must be applied. This normally takes the form of limiting the estimated maximum stress to something well below the hull material's yield strength (see next chapter).

9.5 Optimum distribution of longitudinal structure. Figure 12.9 shows how both hogging and sagging bending moments vary over the length of the ship. A simple knuckled three-segment envelope is shown as a sort of guide to longitudinal intensity of applied moments. As a general rule, it can be seen that the maximum moments are likely to occur somewhere within the midship half length and to drop down to zero at bow and stern. These facts are recognized by the naval architect in working out the details of structural design. The longitudinal structural material is made heaviest through the mid-length and tapered toward the ends. It does not reduce to paper-thin material at the ends, however, because local loads predominate in those locations. The lower diagram in Fig. 12.9 shows how the strength of structural components should vary from bow to stern.

From all of the above, we come to the following conclusions as how to distribute a ship's structural material so as best to resist longitudinal bending moments:

(1) Concentrate the strongest (that is, thickest) materials in those parts of the structure that are farthest from the neutral axis: the uppermost continuous deck and bottom shell.
(2) Place the strongest (that is, thickest) material in those components that run continuously fore and aft.
(3) Maintain the maximum standard of strength throughout approximately the midship half-length, and then taper down (but not to zero) at the ends.

Further Reading

Paulling, J. Randolph, "Strength of Ships," Chapter IV in *Principles of Naval Architecture,* Edward V. Lewis, Ed., Society of Naval Architects and Marine Engineers, Jersey City, N.J., 1989.

Beer, Ferdinand P. and Johnston, Russell, *Mechanics of Materials,* McGraw-Hill, New York, 1981.

Fig. 12.10 A mistreated little ship. This Alaskan passenger ship had the misfortune to be run aground during high tide. Ships are not designed for such treatment, and its bottom was badly damaged. In this case the ship was only lightly loaded, otherwise it might have broken in two.

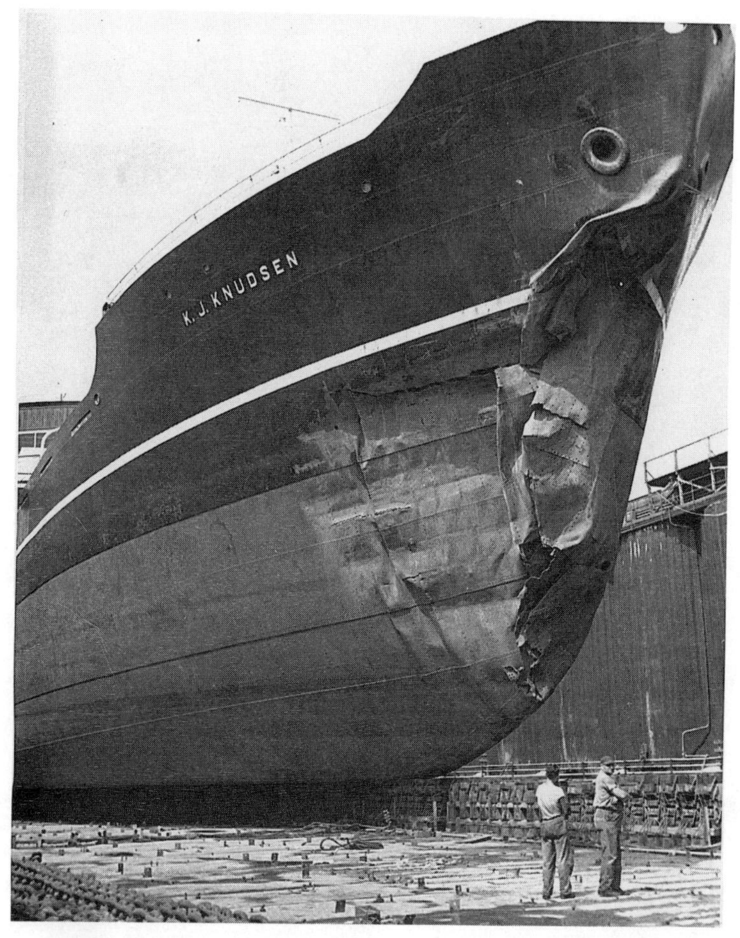

Fig. 12.11 Another mistreated ship. This shows damage to the striking ship in a collision. Ship structures are not intended for such incidents, but a collision bulkhead near the bow confines the flooding to the forepeak tank. Courtesy Newport News Shipbuilding and Dry Dock Co.

Fig. 12.12 Two wartime hull failures. Upper: Tanker *Robert C. Tuttle,* which broke in two while at anchor owing to a combination of previous torpedo damage and some thoughtless pumping of ballast. Lower: Tanker *E. H. Blum,* which broke in two after striking a mine. This shows brittle fracture. Courtesy Newport News Shipbuilding and Dry Dock Co.

CHAPTER XIII

MATERIALS OF CONSTRUCTION

1. *Preface*

Ah! what a wondrous thing it is
  To note how many wheels of toil
One thought, one word, can set in motion!
  There's not a ship that sails the ocean,
But every climate, every soil,
  Must bring its tribute, great or small,
And help to build the wooden wall!

(from LONGFELLOW's "The Building of the Ship")

As Longfellow implied, the variety of materials in a vessel, no matter what its size, is truly impressive. This is especially true in the outfitting category. In the interests of keeping this book down to reasonable size, however, I shall confine my exposition to the principal materials currently used in hull construction. Although much is left unsaid, a foundation is laid on which you may build by studying the sources cited at the end of the chapter.

Technical improvements in boats and ships follow improvements in materials of construction. For example, hydrofoil boats were tried by Alexander Graham Bell back around the turn of the century, but never became practical until lightweight engines and hull materials became available. Naval architects and marine engineers always try to stay abreast of what is going on in material science and engineering. New developments in engineering materials continually invite new possibilities in design.

2. *Physical Properties*

2.1 Fabricating and joining. When the naval architect is choosing a material from which to build a hull, the question of ease of construction should be paramount. Can the material be easily formed in two dimensions? In three? If three-dimensional shaping is difficult, can the vessel be redesigned to allow, for example, chine (versus round-bottom) construction? Most kinds of materials come in limited lengths and widths. The question then arises: how does one go about joining the discrete parts?

2.2 Tensile properties. Here I need to define three terms. If a steadily increasing tensile load (hence stress) is applied to a structure, the degree of stretching (strain) will at first be in proportion to the increasing load. This will go on until the *proportional limit* (also called *elastic limit*) is reached. If the proportional limit is not exceeded, the material will return to its original length when the load is removed. In certain materials the same may be true at a stress slightly higher than the proportional limit. That is, as a somewhat higher stress is applied, the material will stretch at a faster rate than that of the applied load, but it will still return to its original length when the load is taken off. This slightly higher limit is called the *yield stress.* For many practical purposes, the yield stress and proportional limit may be assumed to be equal.

Now, having reached the yield stress, what happens if the load is gradually increased even more? If metal, the material will now begin to "neck down" and, as a result, will stretch fairly rapidly. That will go on until the material pulls apart. The unit load that produces such failure is called the *ultimate stress.*

While knowing the ultimate strength of a material is of interest, the important thing in design is to be sure that the yield stress is not exceeded.

2.3 Toughness. How well can the material withstand shock? Toughness, the opposite of brittleness, is important in most applications. This was brought out in the previous chapter under the heading of *brittle fracture.*

2.4 Stiffness. Materials that are overstressed in compression seldom fail by simply crushing. What is more likely is that the structure will buckle. Stiffness is a virtue in resisting such failure.

2.5 Bending. In many components of a hull, stiffness is enhanced by bending a few inches of the edge of the material at a 90-degree angle into what is called a *flange.* Some materials will crack if so treated. In the case of steel plates, for example, special "flanging quality" steel is usually required for such components.

2.6 Durability. Corrosion is a never-ending problem, particularly for vessels that spend their lives in salt water. Some materials are far better than others in resisting such attacks; some kinds of wood resist rotting better than others. Continuing improvements in protective coatings offer additional options when selecting hull materials.

2.7 Density. Weight saving is an important consideration in almost any kind of boat or ship. It is of particular importance in high-speed displacement type craft, hydrofoil craft, and air-cushion vehicles. Lightweight structural materials are almost a necessity in such vessels.

2.8 Other physical properties. There are many other characteristics that should be considered in the selection of a hull material. Is the material suitable for mass production? Can it retain its strength in extremely high temperatures? Does it become exceedingly brittle at low temperatures? (Important for ice operations or refrigerated cargoes.) Can it resist continual changes in loading without fatigue? Is it flammable? What are its heat insulating properties? Sound insulating properties? Are its strength characteristics uniform (as in steel) or subject to fairly wide variations (as in any species of wood)? Does it have any inherent ability to resist barnacles and other kinds of fouling? If used with or near other materials will problems, such as electrolysis, occur? Is it nonmagnetic (of interest in minesweepers)? How is it in resisting abrasion? Is it easily repaired, perhaps in remote parts of the world? Can it resist loading normal to the surface? In passenger ships or yachts, is it pleasing to sight and smell? And, the inevitable question that fixes most decisions: how do its costs for material and labor compare with those of alternative materials?

3. Nonmetallic Materials

3.1 Wood. Without question, when it comes to visual appeal, no other boatbuilding material can approach wood. It comes in two major forms: lumber and plywood. Wood, in lumber form, was for centuries the standard material for shipbuilding and, being easily formed, is still popular throughout the world for vessels of limited size.

One of its historic shortcomings was the practical difficulty of making efficient end connections. In former times this technical fact tended to limit boat lengths to the length of available lumber. Greater lengths were possible, but always at a price in strength and complexity. This is no longer true. Modern fastening devices and bonding materials, such as epoxy resin or resorcinol, have freed wooden hulls from the constraints of the past. Joint efficiency, which in bygone days was perhaps only 25 percent, may now be as high as 85 percent. Still, we do not see ships of any great size being built of wood because, even at best, wooden hulls become relatively heavy and expensive as size increases. Moreover, as the world's supply of timber continues to shrink, the availability of suitable lumber becomes a worsening problem.

Another drawback of wood is that, except for a few species, it is subject to rot. In warmer climes it may become infested by an insidious kind of worm called the teredo. Within any given species of timber, one may expect to find wide variations in physical properties, which make it difficult to select scantlings with any degree of confidence. Wood has excellent sound and heat-insulating properties, but it also burns.

Marine-grade plywood is popular in small craft, providing a strong, compact surface for deck, shell, and bulkheads. In using plywood, naval architects must confine hull forms to what are called *developable surfaces.* Those are surfaces that are flat, conical, or cylindrical. Try bending a sheet of paper into a three-dimensional surface and you will see what is meant. The technique of designing a developable surface is well understood, and the task is not difficult, provided knuckles (or chines) are acceptable. In practice, the commercially available length of plywood is often a limiting factor.

The two problems of restricted curvature and limited length can be overcome if the boat is built up of multiple layers of wood veneer hand-glued one over another. That form of construction is labor-intensive and requires some kind of previously prepared three-dimensional mold to provide the desired shape. If properly done, however, this procedure provides a hull of exceptional strength-to-weight ratio.

3.2 Reinforced plastics. You probably don't need to be told that most modern yachts, large or small, are built of some form of fiber-reinforced plastic (FRP). By far the most common variety is glass-reinforced plastic, commonly called *fiberglass* or *GRP.* The main strength comes from the glass filaments, which are bonded by epoxy, polyester resin, or other kind of plastic material. The individual glass filaments are far stronger than steel. They come in three principal forms: ordinary randomly oriented mat, loosely woven braids, and neatly woven cloth. The cloth is the most expensive but is also the strongest (in all directions). The woven braids are next most expensive and offer good strength, but principally in only one direction. In larger boats, panels of the hull structure may be stiffened using sandwich construction. Typically, two layers of GRP are held apart by some stiff, lightweight core material such as styrofoam or balsa wood.

Where weight saving is highly important, fibers other than glass may be used. Carbon fiber is one example.

Building an FRP boat involves fairly simple hand labor. The builder cuts sheets of glass fibers (or whatever) to a convenient size and shape. These are laid over, or into, a three-dimensional form (called the *mold*) and the plastic, in liquid form, is then spread evenly over the sheets. Then additional sheets and bonding material are added in turn until sufficient thickness is attained. Usually the most difficult task is in providing the mold. Sometimes this is taken care of by using an existing boat for that purpose. For top-quality work, however, it is better to use a concave mold. That is more likely to lead to a smooth exterior in the finished product. The usual procedure for making a concave mold is first to carve a convex form and use that as a plug to cast the working mold. All this involves a good deal of hand labor and expense, which suggests the wisdom of spreading the initial cost over many identical hulls made from the same mold.

Reinforced plastics are rightfully popular. They provide strong, tough, and attractive hulls, usually at reasonable cost. With even a minimum of maintenance they will far outlast wooden hulls. There are no technical limits on the size of GRP hulls, but few are built over 75 ft in length.

Contrary to popular belief, FRP hulls *will* burn.

Some small craft are built of wood protected by a sheath of clear fiberglass. A cedar-strip canoe covered with GRP is a work of art, bringing delight to the eye and maintenance-free pleasure to the owner.

3.3 Concrete. Concrete is a composite material made of cement (usually Portland) and aggregates such as gravel to lower the cost. It is corrosion resistant and strong in compression. The weakness of concrete in tension may be overcome by incorporating steel rods within the casting. Ships of modest size have been so constructed, but they are found to be uneconomically heavy for cargo transport service. In The People's Republic of China, however, you may see many river barges built of this material.

A special variation is *ferrocement.* Vessels built of ferrocement have relatively thin cement hulls reinforced with a closely spaced steel mesh. This combination is surprisingly flexible and tough, modest in weight, and highly resistant to corrosion. The material is well suited to such applications as commercial fishing boats of intermediate size. There is no reason why ferrocement cannot be used in pleasure craft, too. On the other hand, if you aspire to owning the best-looking hull in the marina, you should look to some other material.

The construction procedure for a ferrocement hull usually involves first setting up a family of bent-to-shape steel bars. These bars, when set in position, give definition to the hull form. Smaller steel bars are then woven into place, and over these is laid and fastened a light steel mesh. The latter may be made of "expanded metal" or perhaps several layers of chicken wire. This metal skeleton is called the *armature.* It not only gives the form to the hull but also provides its tensile strength. With all that steel reinforcement in place, the builders trowel on a thin layer of Portland cement, perhaps premixed with other reinforcement such as whiskers of glass fiber. The builders work in pairs, one inside the boat, the other outside. They apply their trowels in opposition so as to ensure intimate contact between cement and steel. All this should be done in one continuous operation; otherwise, a line of weakness will be inevitable. After this comes the curing period, during which temperature and humidity should be properly controlled.

4. *Steel*

4.1 Manufacturing process. Steel that finds its way into ships may be manufactured by any one of three ways: open hearth, basic oxygen, or electric furnace. The material starts out as a red-hot ingot that is forced into plates or shapes (such as I-beams or angles) by being squeezed between massive rollers while still red hot. Ingots tend to have internal cavities that develop as the material cools. When the ingot is subsequently rolled into useful form, these flaws lead to laminar planes of weakness that may or may not be serious, depending on the application. Where it is necessary to minimize these flaws, special heat treatments and slightly different chemical additives are employed.

Most of the hull structure of a vessel is made up of plates stiffened by rolled shapes. Castings are also used for special complex parts such as the stern frame. Forgings may be used for relatively simple shapes such as a stem bar.

Steel is by far the predominant material used in ship construction today. There are no upper or lower limits on size, although other materials tend to replace steel in vessels under, say, 50 ft in length.

4.2 Standards. In selecting the steel for a ship, naval architects almost always make their selection from some family of recognized standard specifications that prescribe the chemical content and heat treatment. In merchant ship construction, these standards are developed and published by the classification societies (see Chapter I, Section 3.) In the United States, the predominant classification society is the American Bureau of Shipping (ABS). Another prominent set of rules is published by the American Society for Testing and Materials (ASTM). For naval vessels, naval architects must adhere to a set of military specifications, commonly referred to as "the Milspecs."

4.3 Ordinary hull steel. Most of the structure of a merchant ship is ordinary-strength, low-carbon steel. The American Bureau of Shipping offers a menu of six grades: A, B, D, E, CS, and DS. Each grade has a somewhat different chemical content (hence toughness) and some have special heat treatment while others do not, making them suitable for different applications. What they have in common is a uniform yield stress of 34 000 pounds per square inch. Overseas classification societies offer similar selections.

4.4 Higher strength steels. Where greater strength-to-weight ratio is important, classification societies allow the use of a limited selection of higher-strength steels. ABS, for example, permits a special grade with a yield stress of 45 500 pounds per square inch, and another of 51 000 pounds per square inch.

4.5 Special steels. Where special applications place severe demands on steel, special varieties may be required by the classification societies. Liquefied-gas carriers must use a variety of steel that will not become unduly brittle if subjected to extremely low temperatures. The configuration of offshore platforms is such that high tensile loads are often imposed at right angles to plate surfaces, so special attention must be paid to the elimination of laminar lines of weakness in critical parts of those structures. Tankers designed to carry corrosive liquids need special corrosion-resistant steels. Other services may demand steels with high resistance to abrasion.

4.6 Fabricating. Steel plates are usually cut to desired shape by gas cutting or the plasma arc process. Shearing may be used for straight edges. Cylindrical plates are formed by cold-rolling between heavy, adjustable rollers. Three-dimensional shapes can be produced in various ways. For modest degrees of curvature, systematic spot-heating and cooling may be used. For greater degrees of curvature, the plate may be brought up to red heat in a furnace and then forced down over a temporary skeleton-like framework tailored to the desired shape.

4.7 Welding. The individual components of steel hulls are, for the most part, joined together by electric-arc welding. This is done either by hand or by machine. In hand welding, an electric current passes through a specially coated steel rod and makes an arc between the end of the rod and the two parts of the hull to be joined. The heat of the arc melts, in a controlled way, both the hull parts (called the *parent metal*) and the rod itself (called the *filler metal*). The deposited metal fuses into the parent metal and forms a bond between the two parts. As it is heated, the coating on the rod forms a dense gas that covers the molten metal and shields it from the air. Without that shield the molten metal would react with the oxygen in the air and so lose strength. When properly done, a welded joint is practically as strong as the parent metal, but it must be noted that properly welded joints require workers with considerable skill and dedication.

Machine welding is largely confined to long, straight, horizontal joints. In this procedure the welding machine runs along a portable track laid parallel to the joint. The filler metal is carried in the form of a coil of wire and is fed at the proper rate into the joint. Once again, the electric current passing through the rod melts rod and parent metal, causing them to fuse. Air is kept away by a layer of reusable grit that is deposited over the molten metal. Whereas hand welding might require several layers of weld metal, machine welding usually does the job in a single pass. So, machine welding is not only faster, but also more uniform and less likely to contain invisible flaws.

5. Aluminum Alloys

5.1 Classes. Aluminum alloys suitable for marine applications come in large numbers. All can be divided into two categories: heat-treatable, and non-heat-treatable. The former gain much of their strength through the heat treatment. It follows that they are likely to be seriously weakened in the region of any welded joints. The non-heat-treatable alloys gain much of their strength through cold working and retain their strength when welded. These non-heat-treatable alloys are of wider application in hull structures.

5.2 Strength. Among the non-heat-treatable alloys are some containing small amounts of magnesium and manganese. Typical of these is one that has a minimum yield stress of 31 000 pounds per square inch. Another has a minimum yield stress of 34 000 pounds per square inch. Within the group of heat-treatable alloys are several with minimum yield stresses of 35 000 pounds per square inch. As you can see, each of these alloys offers tensile strengths comparable to that of ordinary shipbuilding steel. There are, in addition, many other alloys available, each with its own unique set of characteristics and each with some potential niche in marine application.

5.3 Other important properties. The most outstanding property of aluminum alloys is their low density, which is only about one-third that of steel. You might conclude that a boat built of aluminum would therefore weigh only one-third that of one built of steel. Life is not so simple. In Chapter XII, if you will recall, I mentioned that aluminum's modulus of elasticity is also only about one-third that of steel. Its resistance to buckling failure is therefore much less than that of steel. In vessels of comparable size, as it turns out, aluminum hulls weigh approximately half those of steel.

Another important attribute of aluminum alloys is their high material cost relative to that of steel. This has been the major deterrent to their coming into widespread general use in large ships. In applications where light weight is of overriding importance, aluminum hulls have earned a rightful place. Air-cushion vehicles, hydrofoil craft, and other high-performance craft all make considerable use of the various alloys. As with steel, there are no upper or lower limits on the size of vessels built of aluminum alloys. The weight saving advantages of aluminum, however, are most likely to overcome the cost disadvantage (as against steel) in vessels of limited size, perhaps no greater than 150 ft in length.

5.4 Miscellaneous properties and applications. Aluminum alloys are particularly good in resisting corrosion and are much better than steel in extremely cold (cryogenic) applications such as cargo tanks in liquefied-gas carriers. These alloys have a serious shortcoming in fire-resistant properties and must be heavily insulated wherever danger of fire exists. The low modulus of elasticity can be a problem in some applications, but a benefit in others. It is a valuable property when used in long deckhouses, such as one finds on passenger ships. This is because, being well removed from the neutral axis, long deckhouses are subject to high stresses as the ship hogs or sags. When steel is used, naval architects aim to keep stresses down by resorting to such devices as expansion joints, which bring problems of their own. When aluminum is used, however, its modulus of elasticity (being only one-third that of steel) results in stresses that are only one-third of those

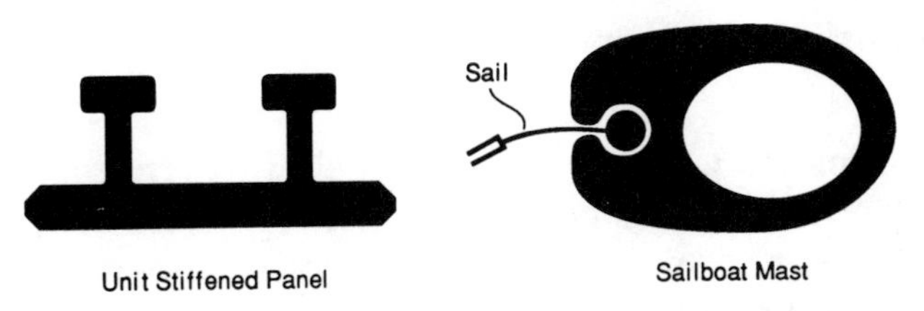

Fig. 13.1 Typical aluminum alloy extrusions.

that would be imposed on a steel deckhouse in the same location. The light weight of the aluminum is also usually of benefit in giving the ship sufficient stability despite the high deckhouse location.

Whereas steel structures must be built up of plates and shapes, aluminum alloys allow complex combinations of plates and shapes to be produced by extrusion. These are limited in their transverse dimensions, but they can be made as long as needed. The extrusion process is used in forming masts for sailing yachts. These not only are hollow but incorporate a slot to accommodate the leading edge of the sail (Fig. 13.1).

5.5 Fabrication and welding. Aluminum alloy plates are usually cut to shape by the plasma arc process, but shearing and sawing may also be used. Aluminum components are welded together, for the most part, using consumable electrodes surrounded by inert gas of some sort. The material cannot readily be welded to steel other than by brazing. Where an aluminum alloy deckhouse is to be fastened to the steel deck of a hull, an arrangement of matching flat bars, such as shown in Fig. 13.2, may be used. The two flat bars can be firmly married by a process of explosive bonding.

Three-dimensional, curved surfaces are far easier to produce in aluminum than in steel.

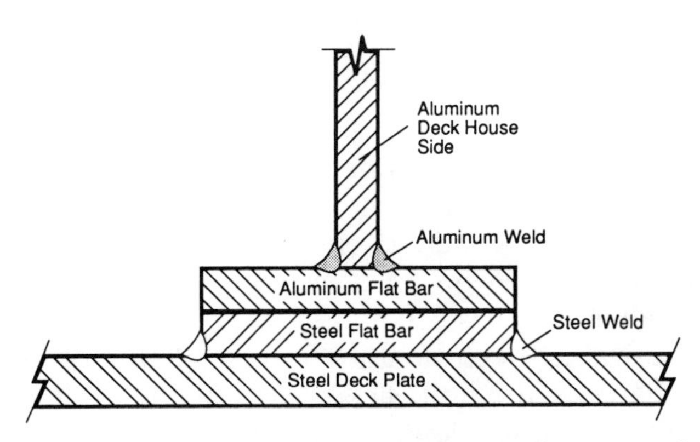

Fig. 13.2 One way to join aluminum alloy to steel (the two flat bars are firmly bonded by an explosive technique.)

6. *Copper-Nickel Alloys*

6.1 Composition. The copper-nickel alloys that have been tried in marine applications are composed principally of copper, with 5 to 15 percent nickel and smaller amounts of other metals. A typical yield strength is 15 000 pounds per square inch.

6.2 Unique properties. These materials offer the appealing advantage of effectively resisting both corrosion and fouling, the twin bugaboos of steel hulls in salt water. Their density is about 10 percent greater than that of steel, a not unreasonable margin when you remember that steel structures need an increment in thickness for eventual wasting away, whereas copper-nickel does not. Material costs are higher than those for steel, but not so much that the material can be dismissed out of hand.

6.3 Applications. Copper-nickel alloys can be applied in two different forms. The simple way is to build the hull out of plates of the alloy, stiffened by steel framing. The more elaborate way is to build the hull of ordinary steel but then sheath it with a thin layer of the copper-nickel alloy. This is done by welding all around the edges of each piece of sheathing material and then "quilting" the remaining surface with *plug welds.* These consist of small (perhaps two-inch-long) slots spaced about six or eight inches apart and filled with weld metal that penetrates into the steel structure below. This sheathing method involves more labor, but saves on material costs. Both methods have been tried in commercial fishing boats of modest size, apparently with considerable success. The sheathing method is more appropriate in larger ships.

There appear to be no insurmountable difficulties in forming or welding these materials. If continuing experience proves the worth of copper-nickel alloys, we may expect to see considerably wider applications in a great variety of vessels and over a broad range of sizes.

7. *Electrolysis*

Naval architects must always be aware of the possibility of interactive damage to hull materials when dissimilar metals are immersed in salt water. Salt water conducts electricity. As such, it serves as an electrolyte, helping to set up an electrolytic reaction between, say, a bronze propeller and a steel hull. Whereas electroplating of old silverware is desirable, electroplating the propeller with steel from the hull is rather hard on the hull. There are various ways to overcome the problem. The first is to avoid, as much as possible, using dissimilar metals in salt water. Where that is impractical, two different techniques are available for minimizing electrolytic damage. One is to attach to the hull a number of *sacrificial anodes.* These are small pieces of some material, such as aluminum, zinc, or magnesium, that will waste away more readily than the metal they are intended to protect. As the name implies, they must be replaced periodically. This system is often used, not only on hull exteriors, but also in the cargo holds of tankers. Another approach provides cathodic protection through an electric current impressed on the hull.

Further Reading

Stern, Irving L., "Hull Materials and Welding," Chapter VIII in *Ship Design and Construction,* Robert Taggart, Ed., Society of Naval Architects and Marine Engineers, Jersey City, N.J., 1980.

Hamlin, Cyrus, *Preliminary Design of Boats and Ships,* Cornell Maritime Press, Centreville, Md., 1989.

Devoluy, Raymond P., "Hull Preservation and Maintenance," Chapter XIV in *Ship Design and Construction,* Amelio D'Arcangelo, Ed., Society of Naval Architects and Marine Engineers, Jersey City, N.J., 1969.

CHAPTER XIV

STRUCTURAL ARRANGEMENTS

1. *Intent*

In this chapter my first aim is to introduce you to certain key, practical considerations that govern a naval architect's thinking when planning the structural arrangements of boats or ships. A second aim is to show how structures are actually arranged in a variety of vessels built of a variety of materials. The third, and final, aim is to define some of the more important technical terms applied to the many different components that make up the hull structure.

When you have absorbed what is in this chapter, you will be able to look at the structure of a boat or ship with some understanding of why the parts are arranged the way they are and what each component contributes to the overall structural integrity. Nothing will be said, however, about the fine points of selecting plate thicknesses or framing sizes (that is, scantlings). Nor will anything be said about predicting stresses within structural components.

2. *Practical Considerations*

2.1 General. In Chapter XII you learned that the main overall (that is, longitudinal) strength of a hull, which keeps it from breaking in two, is supplied primarily by the plating in the uppermost deck and bottom shell. All that plating is held up and forced to do its work by internal stiffening members, the collective term for which is *framing.* Most of what I have to say in this chapter concerns these framing members.

2.2 Self-stiffening. In a minority of cases, what is said above about framing is not true. I refer to plating that is arranged so as to be self-stiffening. A round-bottomed aluminum canoe, because of its shape, may require little or no internal stiffening. In ships, bulkheads and sometimes decks may be corrugated, dished, or fluted, so as to require no stiffeners. Lightly built divisional bulkheads may be stiffened simply by flanging the edge of each plate (see Fig. 14.1). Many years ago a few ships were built in England with two or three deep corrugations running along each side throughout most of the length.

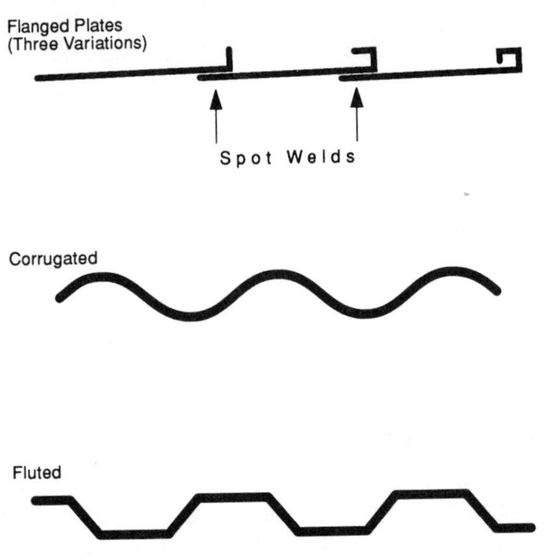

Fig. 14.1 Three kinds of self-stiffened plating.

This reduced the need for framing and supposedly led to appreciable weight saving. The idea, while interesting, has failed to catch on.

2.3 Plates and shapes. The great majority of plating, whether on deck, shell, or bulkhead, is held in position by some form of framing. Most of these framing members in a steel or aluminum vessel are simple standard angles welded in the so-called "inverted" position, as shown in Fig. 14.2. They are called "inverted" because angles were originally turned the other way so that they could be riveted to the

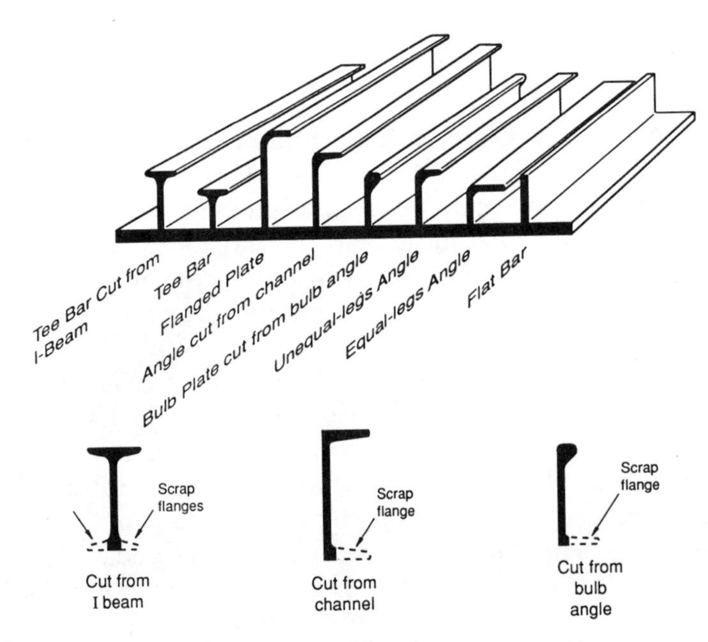

Fig. 14.2 A menu of framing components.

plating. Other framing members include flat bars, bulb plates cut from bulb angles, and T-bars, all as shown in the figure. Where heavier members are needed, angles may be made from channels with one flange removed or, for even deeper frames, flanged plates may be used.

Those members that are produced by removing flanges from standard sections reflect the unhappy fact that American steel mills are still living in the days of riveting.

The scantlings of rolled steel members are specified in certain well-understood patterns. In the case of angles, one simply states the widths of the legs and the thickness. For example: six by four by a half. (That all dimensions are in inches is understood.) For the other rolled sections the web and flange thicknesses are different, so instead of specifying thickness, the scantlings are given in terms of the weight per linear foot of the member. For example: five by three by ten pound I-beam (written $5 \times 3 \times 10\#I$). In the case of steel plates the scantling is commonly given in terms of the weight per square foot. A plate one inch thick and one foot square happens to weigh 40.8 lb. If it were half an inch thick, it would weigh 20.4 lb, and so forth, and that is the way the thickness would be shown on the drawings. In conversation, however, the fractional parts would be dropped and the half-inch plate would be referred to as "twenty-pound plate."

Most framing is oriented in one of two ways. The old, traditional way, dating back to wooden hulls, is to run the framing around the hull in the transverse direction. This is the way your ribs are arranged and explains why a boat's frames are often called its "ribs." An important detail to note is that wherever possible the frames should be continued around the hull in a closed loop. That is, the bottom frames should be in line with the side frames, which should be in line with the deck beams, and so on right around the hull. Transverse framing is almost necessary in planked wooden boats, the frames serving to hold the strakes (planks) together. Transverse framing is also popular in vessels built of other materials, for reasons explained below.

The main alternative to transverse framing is longitudinal framing, with the stiffening members running from bow to stern. If correctly designed and built, this system offers the important advantage of adding directly to the longitudinal strength of the hull. As you may readily understand, before a longitudinally framed ship can break in two, the frames as well as the plates must fail. Longitudinal framing may therefore offer a weight-saving advantage over transverse framing.

Some ships use transverse and longitudinal framing in combination. Some multideck ships, for example, have longitudinal framing in the bottom and at the main deck. The sides, however, are transversely framed.

The reasoning behind this combined framing approach illustrates a couple of important points. One is that framing should as much as possible be oriented across the short dimension of a rectangular area of plating. That keeps the span as small as possible and so minimizes the applied bending moments. In multideck ships, the decks may be about eight feet apart, while the bulkheads (forming the other sides of the rectangle) may be perhaps fifty feet apart. Transverse side frames, running up and down from deck to deck, are the obvious choice. In the bottom structure of the same ship, the rectangular plating areas are bounded by transverse and longitudinal girders within the double bottom. These are probably roughly equal in spacing, so the framing members are run fore and aft so as to add to the longitudinal strength of the hull girder. The same argument applies to the main deck, although the details differ. Keep in mind, too, that the bottom shell and main deck, being well away from the neutral axis, are in the best positions to add to the strength of the hull girder.

Another way to combine framing systems is to use longitudinal framing over most of the length of a ship, but transverse framing at the ends.

A third kind of framing system may be found at bow or stern. At the ends, the hull may be so shaped that diagonal framing makes sense. Such framing fits against the plating at about 90 degrees, lending maximum strength. Figure 14.3 shows two examples of diagonal framing: *cant frames*, typically used in rounded sterns (which, in truth, are

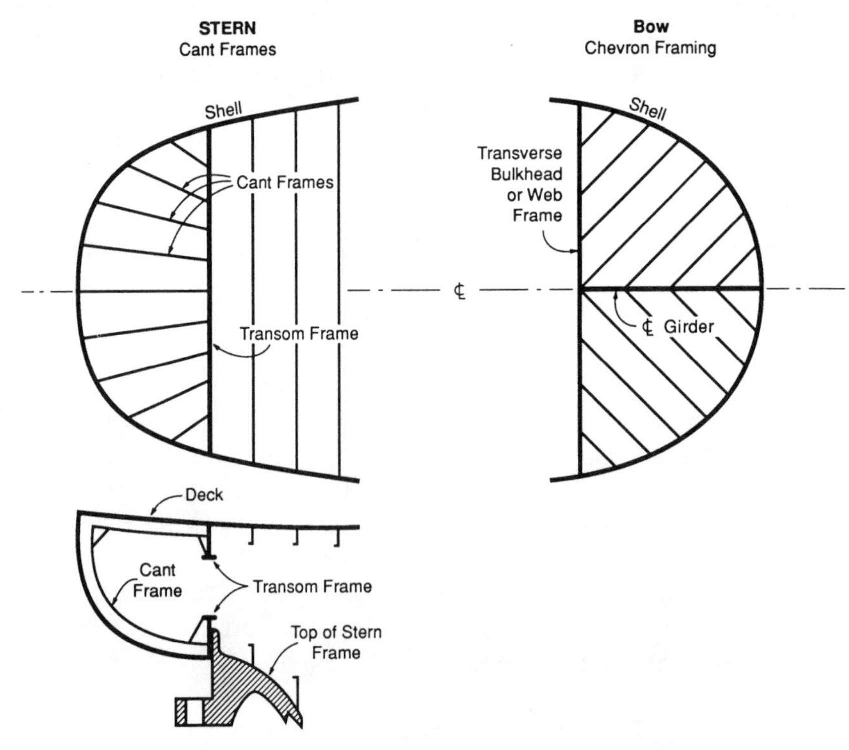

Fig. 14.3 Cant frames at stern and chevron frames at bow.

no longer popular), and *chevron frames,* which you may find in blunt bows, such as those found in Great Lakes bulk carriers. Note that the forward end of the cant frames are supported by an extra-heavy transverse web called the *transom frame.*

A key step in structural design is deciding how far apart to space the frames. For any given set of circumstances, there is some theoretically optimum spacing that will lead to minimum weight. In practice, however, some other spacing is often chosen. Wider-than-optimum spacing leads to fewer parts to be fabricated, fitted, and welded, which leads to reduced labor costs. Wider spacing also gives the welders and other mechanics more room in which to work. In transversely framed ships, the spacing through most of the length is held uniform. At bow and stern, however, because of higher local loads, the spacing may be reduced.

With the frame spacing established, how does the naval architect decide exactly where each transverse frame should go? Where to start? There is only one fore-and-aft location where a frame really must be placed. That spot is somewhere over the center of the stern frame, where an internal extension of that frame allows it to be strongly attached to the transom frame described above (Fig. 14.3). The transom frame, then, becomes the naval architect's starting point for deciding where all the other frames should go. The naval architect begins at that point and lets the previously decided spacing dictate the location of each additional frame.

2.4 Webs. *Webs,* or *web frames,* are extra-strong framing members. The transom frame mentioned above is an example of a web frame. More commonly, web frames are oriented at right angles (and serve to support) the ordinary frames. They may be at mid-length of the ordinary frames, or at any other uniform spacing, and serve to reduce the span of the frames and so reduce the applied bending moments. In certain locations, such as engine rooms, where extra rigidity may be needed, web frames are used, at intervals, in place of ordinary frames. For example, every fourth frame might be a web.

Web frames may consist of flanged plates, or they may be built up of two pieces: a *web* and a thick flat bar, called a *face plate,* welded to it (Fig. 14.4). Where a web frame crosses an ordinary frame, a notch is cut in the web of the web frame to allow the ordinary frame to continue through in one piece. These notches have to be carefully designed and cut so as to avoid sharp corners, which may lead to high stress concentrations and possible cracks.

Webs and *web frames* are generic terms. In certain specific locations they are given special names. A longitudinal web on the side shell is called a *shell stringer.* One under a deck is called a *deck girder.* Those are just two examples. Others will become evident later.

Vessels of any appreciable size are usually fitted with transverse bulkheads. These not only confine flooding, but contribute to struc-

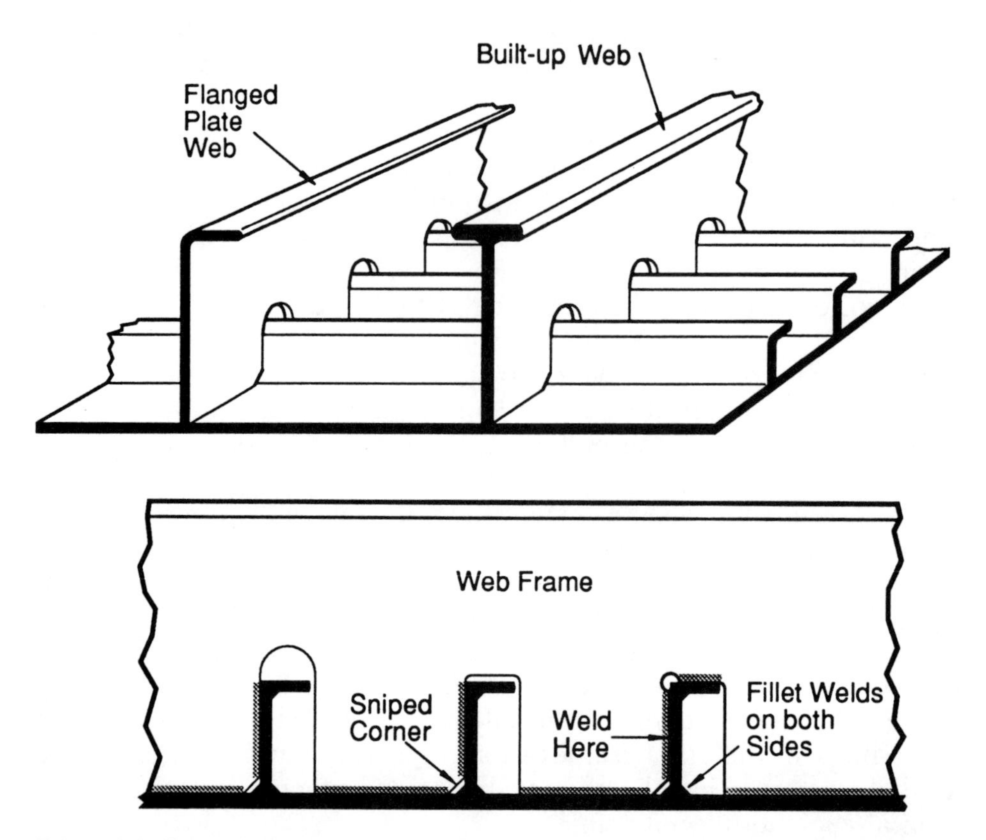

Fig. 14.4 Typical intersections where regular frames pass through web frames. Upper drawing shows alternative web designs (flanged plate and built-up). Lower drawing shows typical notches cut in webs. Rounded corners minimize stress concentrations. Sniped corners leave room for fillet welds joining frames to plating.

tural integrity. They prevent the vessel from racking (that is, losing its transverse rectangularity) and also lend support to any longitudinal framing members. Transverse bulkheads are invariably located on frame lines, which means that each replaces a transverse frame.

2.5 Special demands. Many boats and ships, because of their particular functions, may require special structural arrangements. Stern trawlers, for example, are arranged with a large chute in the stern and carry a massive arched crane (the *gallows*). Both of those features require extraordinary framing arrangements. A *knight's head bow* on a ferry is another example. In Chapter II you learned how cargo access systems make special demands on general arrangements. In wrapping the design around the cargo system, the naval architect may have to pay a lot of attention to special structural arrangements.

2.6 Some miscellaneous considerations. In laying out the structure of a vessel, the naval architect should keep in mind the fitters and welders who must assemble and weld the various parts. In a

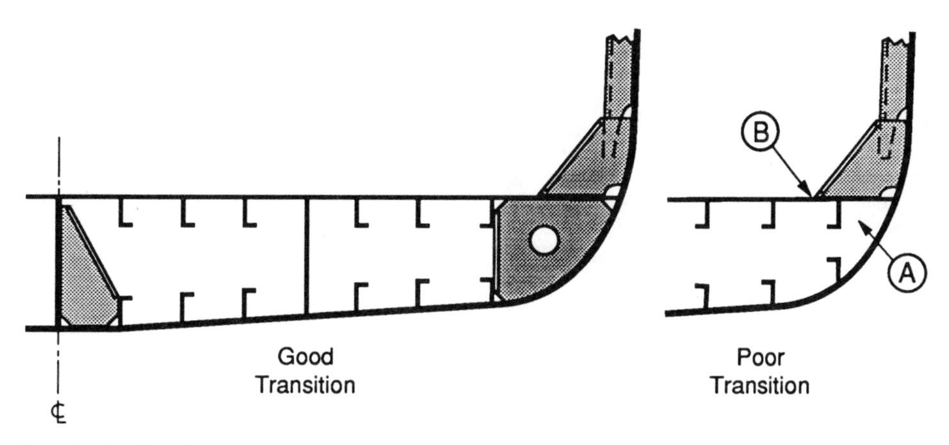

Fig. 14.5 Right way and wrong way to effect a transition between transverse and longitudinal framing. (A) Shows an inaccessible pocket, impossible to fit, weld, or maintain. (B) Shows a "hard spot," sure to produce a crack.

longitudinally framed doublebottom, for example, transverse brackets should replace the outermost longitudinals, as shown in Fig. 14.5. If, as shown, the ship happens to have transverse side framing, those brackets provide an ideal transition between the two kinds of framing.

I explained in Section 2.3 that framing members should, wherever possible, be laid out in complete loops. Where this is not possible, the designer should avoid "hard spots," because such stress-raisers are likely to lead to cracks in the structure. Avoiding hard spots can be done either by using tapered endings or by adding local reinforcement. Figures 14.5 and 14.6 show examples.

3. *Some Typical Structures*

3.1 Historic. Much of our structural terminology has its roots in wooden ship construction as practiced in the days of sail (and still

Fig. 14.6 A "hard spot" and two alternative solutions. (A) Shows end of offending member trimmed back. The tapered ending avoids sudden discontinuity. (B) Shows a flanged bracket, which distributes load to an adjacent framing member. The end of the vertical frame is tapered so as to avoid another "hard spot" and also provide ample area for welding to the bracket.

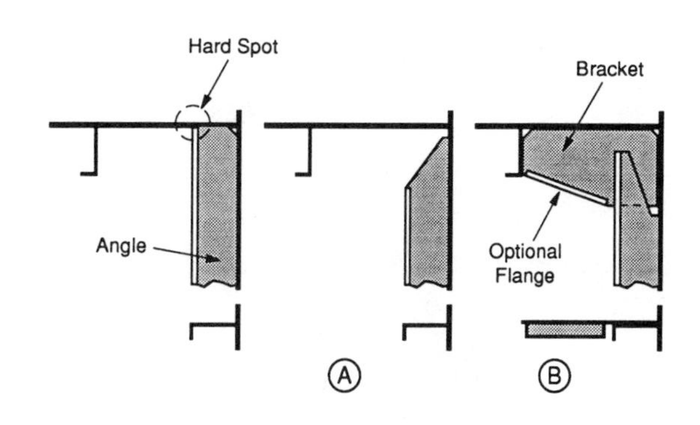

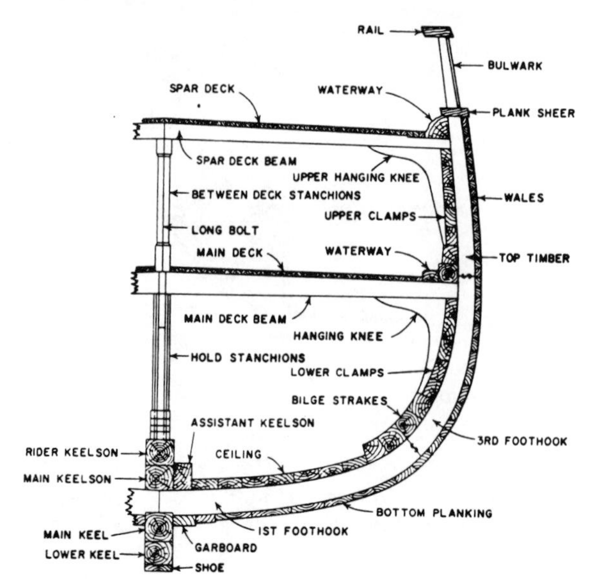

Fig. 14.7 Midship section of a wooden sailing ship (source: Stiansen, 1980).

used in parts of the third world). Figure 14.7 shows the midship section of such a ship. The massive keel and bottom structure reflects the then-prevalent design philosophy that "a ship floats on her bottom and the keel is the backbone of the ship." It was only with the advent of iron that designers realized that the hull should be treated as a giant box girder, which meant that the upper deck was just as important as the bottom shell.

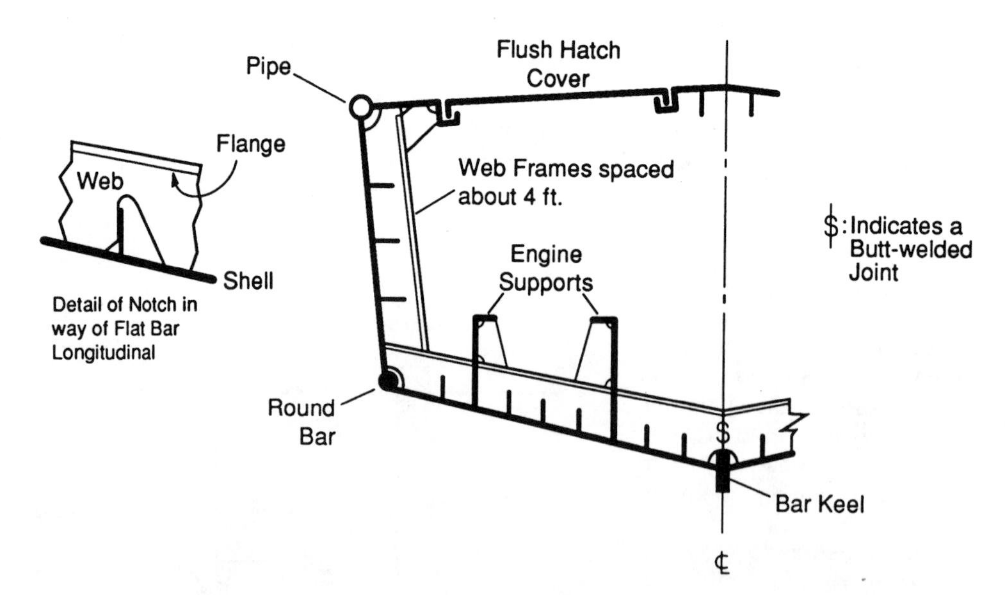

Fig. 14.8 Midship section of a typical small, steel-hulled utility workboat (derived from Graul and Fry, 1967).

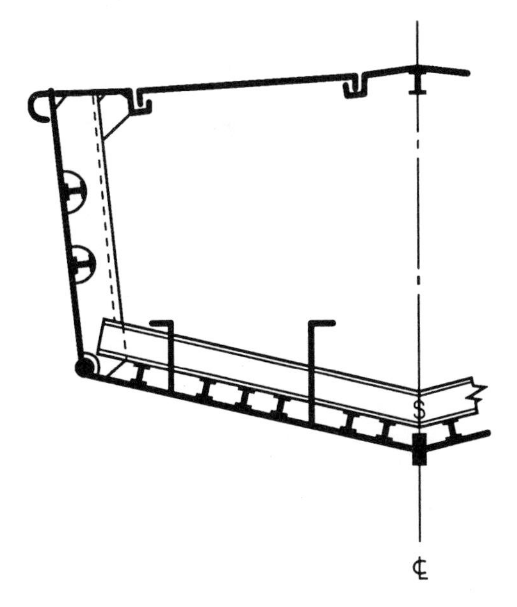

Fig. 14.9 Midship section of a typical, small, aluminum-alloy utility workboat, the equivalent of the steel hull shown in Fig. 14.8. The longitudinals are extruded wide-flange shapes. The bottom web simply sits atop the longitudinals, rather than being notched out (derived from Graul and Fry, 1967).

3.2 Small craft. Figure 14.8 shows the midship section of a typical, small utility workboat with a steel hull. The same sort of construction might be found on pleasure cruisers or recreational fishing boats. The main framing in this case is made up of longitudinal flat bars reinforced with transverse web frames spaced about four feet.

Figure 14.9 shows the midship section for a boat equivalent to that shown in the previous figure, but with an aluminum-alloy hull. In this case the main framing members are extruded "wide-flange" shapes running fore and aft. The bottom web frames are not notched out to accommodate the longitudinals, but simply ride on top of them. This arrangement is not as strong as the notched arrangement, but is considerably easier to put together.

Figure 14.10 shows the midship section for an aluminum-alloy sailboat.

3.3 Small ships. Figure 14.11 shows typical singlebottom construction suitable for vessels of intermediate size such as oceanographic research ships. In this case the ship has transverse framing, and the side shell frames are inverted angles. In the bottom, however, in order to give greater strength, the angles are replaced by flanged plates, called *floors.* Running fore and aft are three girders, each equal in depth to the floors. The one on the centerline is called the *center keelson* (pronounced "KEL-son") or *center vertical keel,* abbreviated CVK. Those on the sides, as you can see, are called *side keelsons.* The term *intercostal* (meaning "between the ribs") tells you that the side keelsons are made in short lengths that fit between the floors. Alterna-

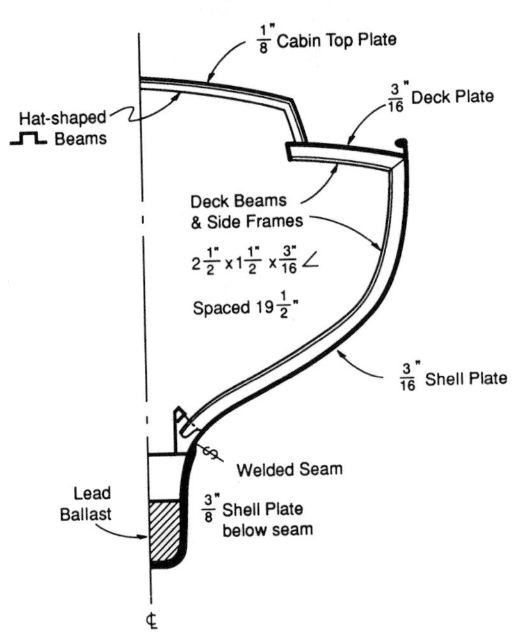

Fig. 14.10 Midship section of a 30-ft *LWL* sailboat of aluminum alloy.

tively, the keelsons could be made continuous and the floors intercostal.

The *limber holes* shown in Fig. 14.11 allow bilge water to drain toward some central location where it can be pumped out.

3.4 Large ships. Figure 14.12 shows the midship section for a large containership. As you may see, the hull is essentially a U-shaped structure with only one deck, and that is mostly cut away so as to ease

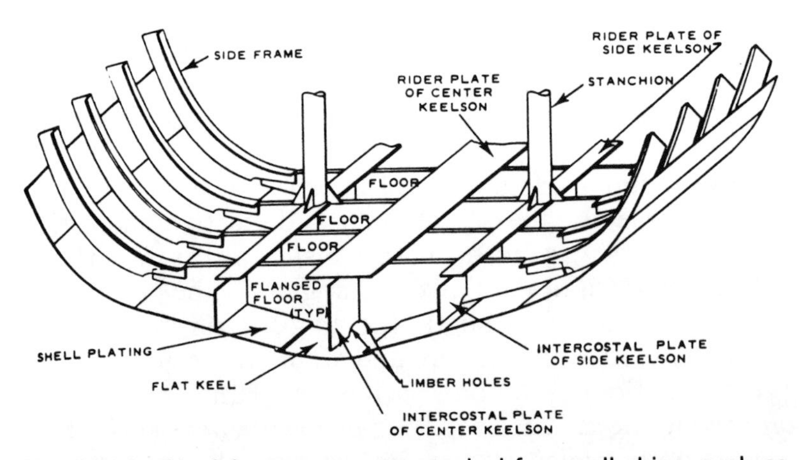

Fig. 14.11 Singlebottom structure typical for small ships, such as tugboats, trawlers, and oceanographic research ships (source: Stiansen, 1980).

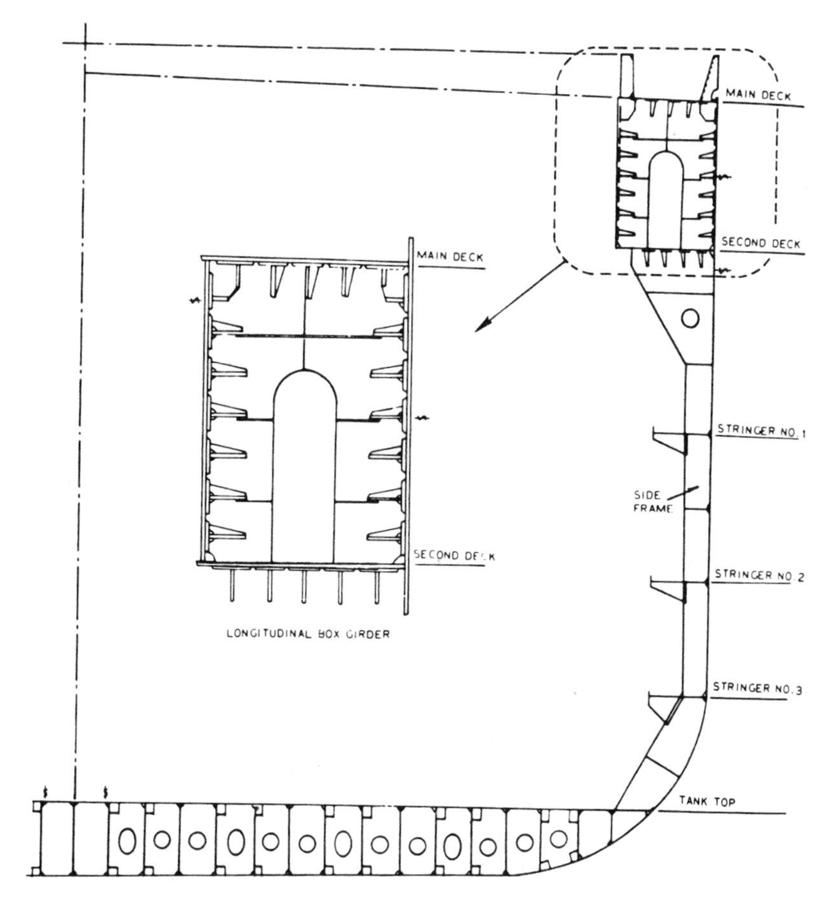

Fig. 14.12 Midship section of a large containership (source: Stiansen, 1980).

the work of moving containers in and out. (The containers are simply piled one atop the other, filling the hold). The deck opening is protected by strong hatch covers and more containers are piled thereon, perhaps four high. To ensure the longitudinal strength of the ship in spite of the "all-hatch" arrangement, massive steel box girders (big enough to walk through) are fitted in the upper outboard corners as shown in the detail. The plating of the girders is supplemented with heavy, closely spaced longitudinal flat bars. The doublebottom is longitudinally framed and the sides, below the corner girders, are transversely framed. The hatch covers are not shown, nor are the vertical guides that keep the containers in vertical alignment, nor the securing devices for the containers piled on the hatches.

Figure 14.13 shows the general arrangements of structure within several widely differing kinds of large merchant ships. The drawing shows only the main fore-and-aft plating elements. The framing members are not shown, but they might be arranged either transversely

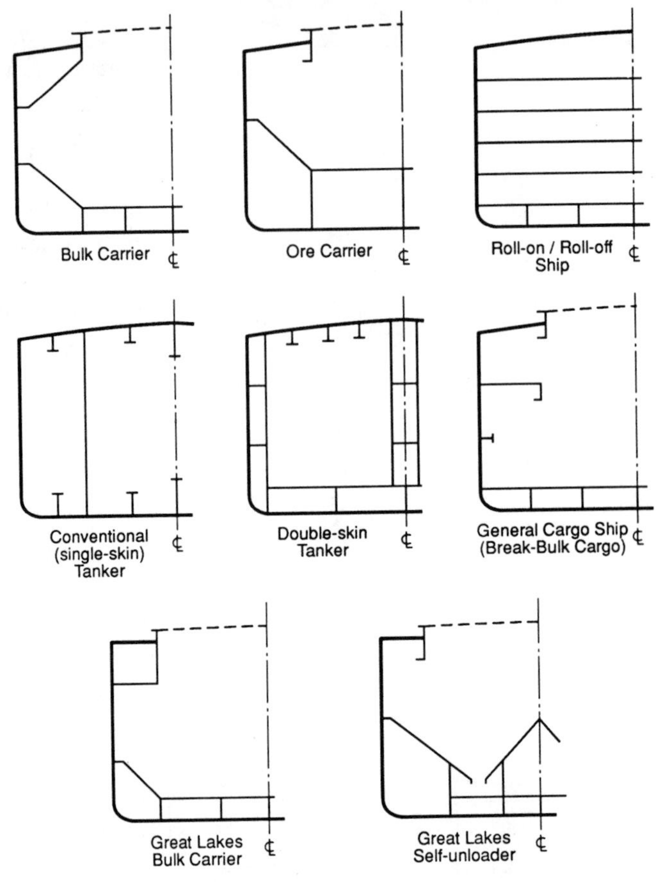

Fig. 14.13 General structural arrangements for a variety of cargo ships.

or longitudinally depending on circumstances. Please note that the individual sections are *not* drawn to the same scale.

References and Further Reading

Stiansen, Stanley, "Structural Components," Chapter VII in *Ship Design and Construction,* Robert Taggart, Ed., Society of Naval Architects and Marine Engineers, Jersey City, N.J., 1980.

Graul, Timothy and Fry, E. D., "The Design and Construction of Metal Planing Boats" in *Proceedings,* Society of Naval Architects and Marine Engineers, Spring Meeting, Jersey City, N.J., 1967, pp. 8-1 to 8-24.

D'Arcangelo, Amelio M., *A Guide to Sound Ship Structures,* Cornell Maritime Press, Centreville, Md., 1964.

Fig. 14.14 *ARCO Alaska* under construction. This 188 000-dwt-capacity tanker features a doublebottom. The picture clearly shows the twin longitudinal bulkheads and longitudinal framing that are typical in tanker structures. Courtesy Peter E. Jaquith, National Steel and Shipbuilding Co.

CHAPTER XV

MISCELLANEOUS DESIGN MATTERS

Preface. How should I bring this book to a close? There are still many important matters you might want to have explained, but where do I draw the line? As a compromise, I present below a few introductory paragraphs on each of half a dozen assorted topics. If these arouse your curiosity, I urge you to dig further in the books and articles recommended at the end of each section.

1. Steering and Maneuvering

1.1 Scope. My intent in this section is to tell you about the problems inherent in making a boat or ship stay on a straight course, or turn, stop, or go astern. My further intent is to give you at least a rough idea about how naval architects design rudders and other mechanical devices to provide enough control to keep these problems in hand.

1.2 Steering. Many floating craft are directionally unstable. By this I mean that when moving ahead through the water, if they are deflected slightly off course, they will tend to swing even further. This comes about because the vessel's turning center will seldom be found more than 20 percent of the vessel's length from the bow, whereas the longitudinal center of gravity is usually somewhere near mid-length. This misalignment provides a turning couple that encourages further deflection off the original course, as shown in Fig. 15.1. There is, moreover, a lateral hydrodynamic force that also tends to increase the swing. This arises because, when the vessel starts to yaw, the asymmetry of the bow wave places a crest on one side of the bow and a hollow on the other (Fig. 15.1). (Remember that although the vessel has pivoted to one side, its momentum is still pushing it along its original course so that it is moving ahead in a crabwise fashion.) If left unattended, the vessel will continue to veer until its centerline coincides with its actual direction of motion. Then any externally applied lateral force may trigger another major swing to one side or the other. The usual way to keep a vessel on a straight course is to use a rudder to offset any external deflecting force.

As implied in the previous paragraph, one of the rudder's main functions is to keep the vessel on course. The other is to make the vessel turn off course. Unfortunately, the physical characteristics of ship and rudder that make turning easy usually also make steering a

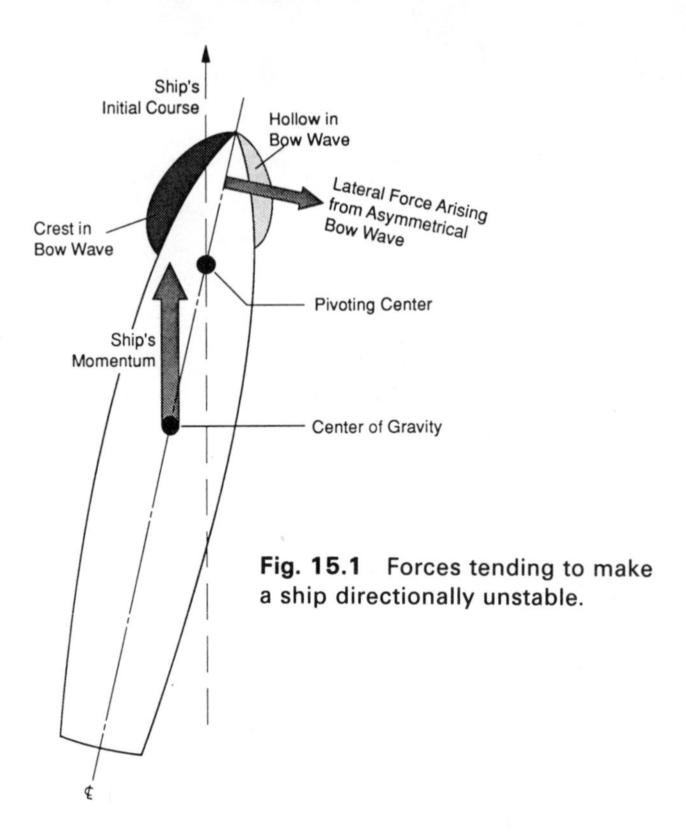

Fig. 15.1 Forces tending to make a ship directionally unstable.

straight course difficult. Compromises must be made. In some kinds of vessels, such as deep-sea merchant ships, or cruising yachts, ease of steering a straight course is the important thing. In combatant naval vessels, racing yachts, or harbor tugs, quick turning takes precedence. In either case, naval architects recognize that rudder action inevitably produces drag; so the challenge is to produce a configuration of rudder and hull that will provide an acceptable degree of steering control with only moderate added resistance.

1.3 Hull form effects. In a vessel where easy course-keeping is important, the design should feature as far as possible a trim aft, a cutaway forefoot, a full deadwood (see Glossary), and fine lines. These characteristics tend to shorten the distance between the vessel's center of gravity and its turning center, which thereby reduces the turning couple that tends to swing the craft off course. These same characteristics lead to large turning circles, but that is the price that must be paid.

One of the greatest challenges to course-keeping arises when a vessel is operating in a following or quartering sea. There is then a strong tendency for the craft to yaw and turn broadside to the waves. Such action may lead to a capsize. The best way to minimize that trouble is to avoid a wide, flat stern and deep forefoot.

1.4 The turning path. Figure 15.2 shows the path a typical vessel will follow when the rudder is put over as far as it will go (usually 35 degrees) and the engine is kept turning at full power. The figure helps define four pertinent terms: *advance, transfer, tactical diameter,* and *steady turning radius.* Please notice what happens right after the rudder is first swung over (identified as "execute position of O" in the drawing). Because the rudder force is directed to port, the vessel will move off slightly in that direction. In short order, however, the rudder action swings the vessel so that its centerline is angled off to starboard relative to its actual direction of motion. Now the hydrodynamic forces on the hull itself take control and put the vessel into a spiral path that soon turns into a steady circular course. Notice that the vessel's centerline is always turned inward from the actual course. It is the hydrodynamic results of this yaw angle, rather than the rudder force, that produces most of the turning effect.

1.5 Rudders. Why are rudders usually located near the stern? In order to maximize the turning moment, the rudder should be as far as possible from the vessel's turning center. As mentioned above, that center is seldom more than 20 percent of the vessel's length from the bow, and that answers our question. It also explains why bow rudders (which are seldom fitted) are effective only when going astern. The effectiveness of a rudder is greatly enhanced if it can be placed immedi-

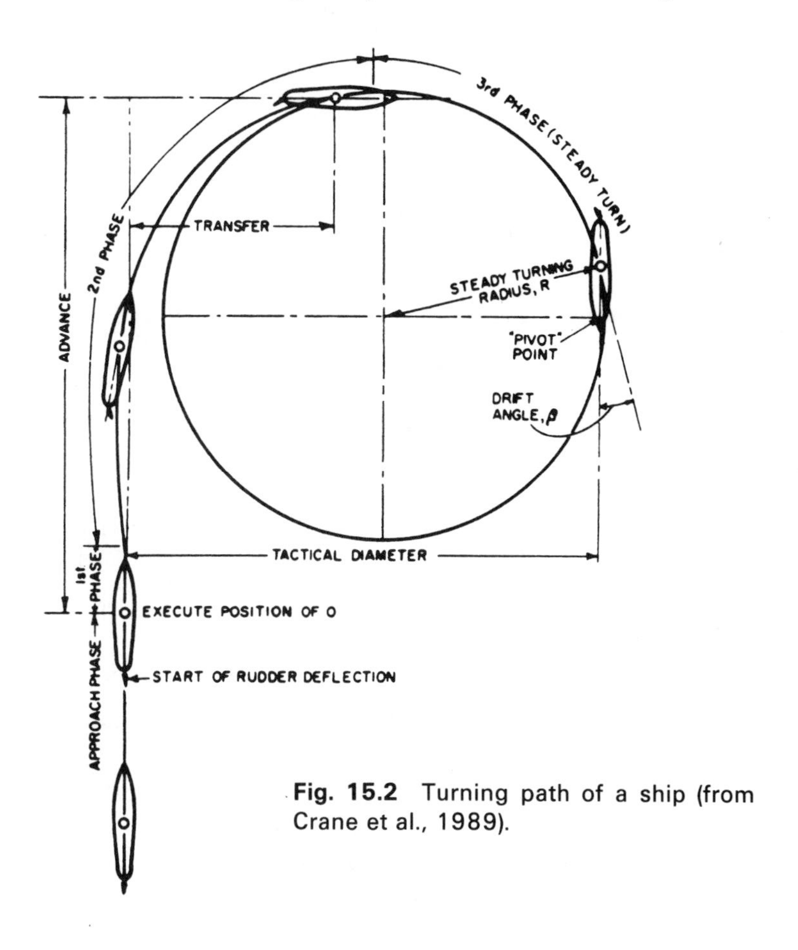

Fig. 15.2 Turning path of a ship (from Crane et al., 1989).

ately aft of a propeller. This is because the impact of the water leaving the propeller will increase the lateral forces exerted by the rudder when it is turned, and can help swing a ship even when it has no headway. On the other hand, twin-screw ships with a single centerline rudder are notoriously sluggish in responding to the helm.

How big should a rudder be? That will depend on how important good maneuverability is on the craft in question. It will also be influenced by the type of rudder (about which more in a moment), and its location. In general, rudder area is chosen relative to the vessel's underwater profile area. In sailing yachts, for example, rudder areas generally fall between 8 and 11 percent of the profile area (with the higher values being appropriate to smaller craft). In larger ships, the usual profile area is almost exactly equal to the product of the vessel's waterline length and draft. Since that product is easier to find than profile area, it usually becomes the basis for choosing rudder area. Typical ratios run from 1.5 percent for large single-screw ships to 5.0 percent for commercial fishing boats. In sailing ships the value may be about 3.0 percent. In pilot boats or ferryboats, where good maneu-

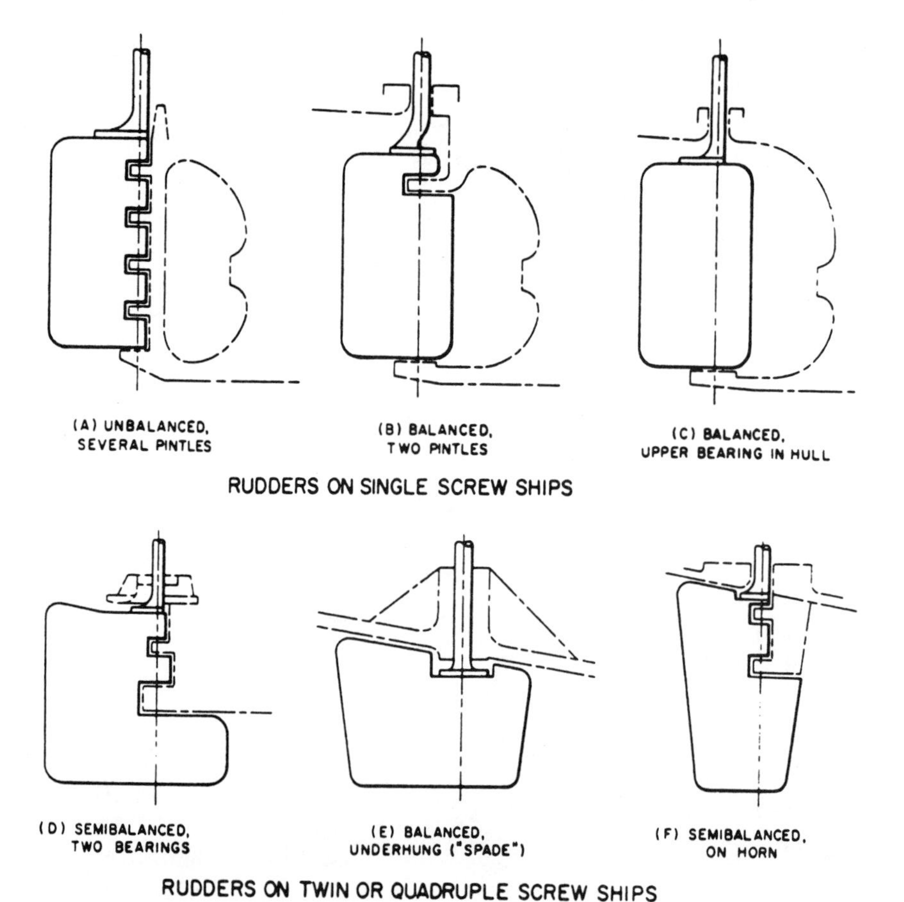

Fig. 15.3 Six conventional kinds of rudders (from Hunley and Lemley, 1980).

verability is important, the figure may easily reach 4.0 percent. Motorboat ratios are typically around 4.0 to 5.0 percent, whereas seagoing tugs usually range from 3.0 to 6.0 percent.

Figure 15.3 shows typical conventional rudders suitable for ships or small commercial single-screw craft. Those with some of their area forward of the stock are called *balanced* rudders. In designing such a rudder, the architect's aim is to place the stock just a little bit forward of the center of lateral pressure exerted on the rudder when it is turned. This center invariably turns out to be well forward of the center of area (just as we have seen to be true of the lateral pressure on the hull itself when the vessel is turning). You will see this reflected in the balanced rudders shown in Fig. 15.3.

If you will refer back to Fig. 11.6 you will see typical rudder arrangements for sailboats. In the lowermost sketch you will notice that the rudder has been separated from the keel. With a short fin keel like that shown, a rudder at the trailing edge would be so close to the boat's turning center that it would be quite ineffective.

In a small craft the rudder may be controlled by a hand-operated lever called the *tiller,* or, in somewhat larger craft by a hand-operated steering wheel. In yet larger vessels some sort of mechanical help will be needed, usually in the form of an electrically powered hydraulic steering engine located within the hull just above the rudder. The engine is remotely controlled from the bridge, but in an emergency it can be controlled *in situ.*

1.6 Going astern. Steering a single-screw vessel while going astern is seldom easy. Because of the usual shape of the stern lines, the swirling water coming out of a backing propeller impinges with greater force on that part of the hull above the shaft than that below. As a result, there is a strong tendency for the stern to be pushed to one side, usually to port. The rudder is remarkably ineffective. All of the factors that made its location so good for steering while going ahead are now working against it as an instrument of control. A few years ago I was involved in operating a "narrow boat" on one of the canals in England. We learned the trick for steering while going astern. Periodically we switched the engine from backing to full ahead for a few seconds. That would allow us to get some momentary turning force out of the rudder and so aim the boat where needed. We also observed an unhappy pair of related facts. One was that the typical canal was just as wide as the boat was long. The other was that, in going astern, the boat's strong natural proclivity was to turn 90 degrees to the centerline of the canal.

1.7 Other maneuvering devices. Where extra maneuverability is desired, there are several mechanical devices available to the naval architect. You may recall the cycloidal propellers mentioned in the chapter on propulsion devices. Twin screws can also help, especially when the vessel is moving ahead at good speed. In harbor maneuvers, however, twin screws are usually painfully slow in effecting an otherwise unassisted turn. A far better device for harbor maneuvers is the side-thruster, which consists of a reversible-direction propeller fitted within a cylindrical duct located just above the keel and running across the ship and open to the sea at each end. The typical thruster is located near the bow, but thrusters may also be placed near the stern. In some cases, two or three thrusters may be fitted at each end. Powered by a diesel engine or electric motor, the thruster can exert a strong lateral force either to port or to starboard, as controlled from the bridge. Whereas twin screws work better when the vessel is making good headway, thrusters work well only at low or zero speeds.

More advanced thrusters are also available. Some, for example, take water in from the bottom of the ship and, by means of mechanical controls, obtain thrust by diverting the flow to either side.

A boat fitted with independently operating side paddle wheels can

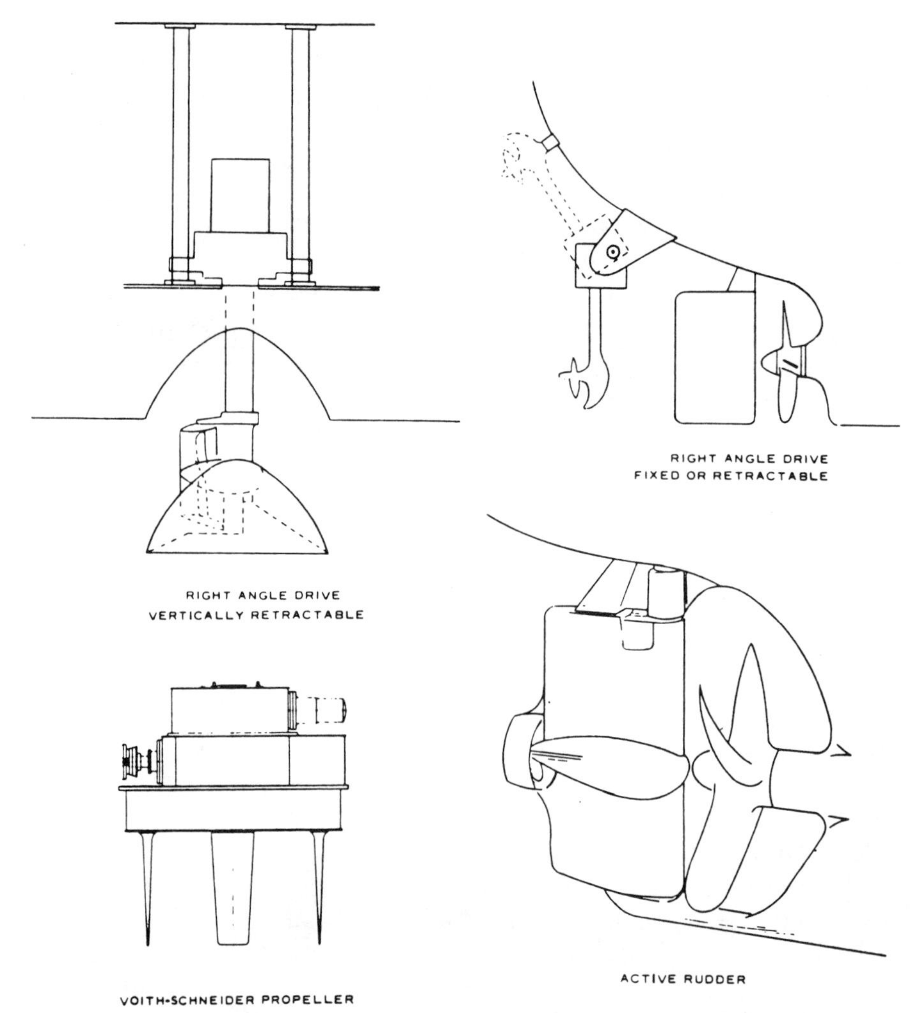

Fig. 15.4 Some maneuvering devices (source: Hunley and Lemley, 1980).

turn on its own axis without trouble. *Quarter wheels* consist of a pair of independent, side-by-side stern wheels. They, too, can be used in opposition so as to turn the boat.

A goodly variety of other maneuvering devices has been developed; Fig. 15.4 shows four typical examples.

1.8 Stopping. Unlike your automobile, ships seldom have brakes. Braking flaps, water parachutes, and other stopping devices are effective only when the ship is moving ahead at high speed. Most braking action, however, is needed in harbors, where the ship is likely to be operating at only about half speed. In practice, then, ship operators rely on running the engine astern to stop the vessel. Because of the time requirements of the method, prudence dictates slow-ahead speed in crowded waters. (The stopping distance, called *headreach,* is closely proportional to the initial ahead speed.)

Out in the open sea, a crash stop from full-ahead speed is usually less wise than throwing the rudder hard over and going into a tight turn. Crash stop tests on some 190 000 DWT tankers showed an average headreach of about 2.8 statute miles. With the turning maneuver, on the other hand, the distance was reduced to about one-fifth that value.

Another influential factor is the ability of the engine to reverse quickly, with the further complication that some engines (notably steam turbines) cannot develop full power in the astern direction. Diesels can develop full power astern, but suffer the disadvantage of being unable to operate in either direction at low speed without stalling. This can be troublesome in harbor operations because of the limited supply of compressed air needed to restart a stopped engine.

Harbor operations are further complicated by the presence of shoal water. The good news is that ships are far easier to stop in shallow water; the bad news is that they are far harder to steer. One series of tests showed that, when moving ahead at half speed, the vessel had a turning circle in shoal water (120 percent of draft) about double that obtained in deep water.

Almost any kind of small craft will be far easier to control than the big ships dealt with above. Small craft have relatively larger rudders that are more quickly turned. Also, they are more easily stopped because their engine powers are vastly greater relative to their displacements. The complicated interactions between all the relevant factors are still there, of course, and an ability to operate the craft safely comes only with hands-on experience.

Further Reading

Crane, C. Lincoln, Eda, Haruzo, and Landsburg, Alexander C., "Controllability," Chapter IX in *Principles of Naval Architecture,* Edward V. Lewis, Ed., Society of Naval Architects and Marine Engineers, Jersey City, N.J., 1989.

Hunley, William H. and Lemley, Norman W., "Ship Maneuvering, Navigation and Motion Control," Chapter XII in *Ship Design and Construction,* Robert Taggart, Ed., Society of Naval Architects and Marine Engineers, Jersey City, N.J., 1980.

2. Vibration

With all of its advantages, we must recognize that mechanical propulsion almost always brings with it problems of vibration. Related to that is the matter of noise, a topic we shall take up at the end of this section. Vibrations are not only unpleasant for the people on board, they may also cause electrical or mechanical equipment to malfunction or fail altogether. The subject is clearly one of great interest to naval architects. They have labored hard and long to develop a theoretical analysis that will allow steps to be taken in design to assure avoidance of unacceptable levels of vibration. Although this effort has achieved some degree of success, there are still cases where the theory leaves something to be desired. Fortunately, if vibration problems arise when the ship is on trials, there are usually practical ways to effect modifications that will cure, or at least minimize, such problems.

The design-stage theory is so complex that it does not form an appropriate subject for a book of this nature. I do, however, want to say something about the sources of shipboard vibration and explain what it is the analysts are trying to determine. Then I shall mention some practical design steps that are usually taken to avoid vibration and also lay out some thoughts on what can be done to make corrections in cases where problems do arise.

1.1 Theory. A ship's hull has several possible modes of vibration. It can vibrate up and down, side to side, fore and aft, or in torsion. In each of those modes, it will have at least one *natural frequency,* just like a tuning fork. That in itself is no problem. Trouble arises, however, when some variable external force is applied with a frequency that coincides with any of the hull structure's natural frequencies. When that occurs, even a small exciting force can lead to major amplitudes of vibration. You see exactly the same effect when pushing someone on a swing. Although each push may be modest, if they are all timed to coincide with the swing's natural period, they soon build up major amplitudes. There is a technical term for this coincidental timing; it is called *resonance.*

The most serious hull vibrations are usually those found in the up-and-down direction. As shown in Fig. 15.5, there can be more than one natural vertical frequency, and these are identified by the number of *nodes* involved. (A *node* is a place in the structure that remains fixed while the rest of the structure is vibrating.) The frequency of vibration

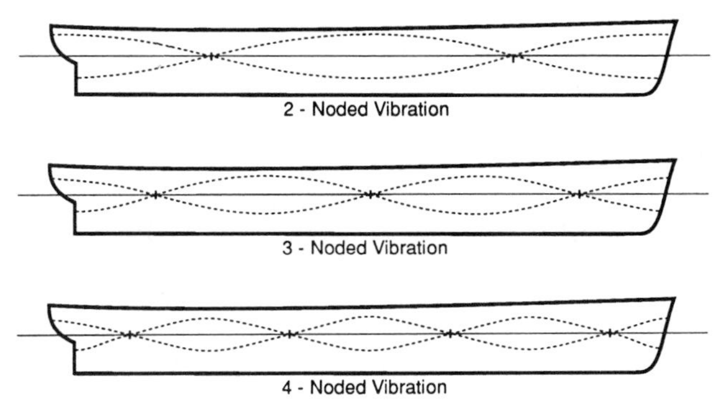

Fig. 15.5 Three modes of vertical hull vibration.

varies with the number of nodes, being higher as the number increases. In most cases, naval architects are primarily concerned with avoiding the two- and three-noded frequencies.

2.2 Sources of exciting force. Some exciting forces may be external to the ship. Pitching and slamming may cause the entire hull to whip, for example. More commonly, however, the sources are internal. An item of reciprocating machinery, such as a diesel engine, may be the culprit. The propeller is the most usual vibration generator, however, for the reason that it seldom operates in a perfectly uniform flow of water. Propeller vibration problems are most apparent in single-screw ships with conventional stern shapes. Here the wake effect is most pronounced on the upper part of the propeller. As each individual blade sweeps around, it continually bites into water coming at it from various directions and at different speeds. The constantly fluctuating reaction force carries through the propeller shaft and imposes a vibration on the hull. Vibratory forces may also be imposed on the hull just above the propeller.

2.3 Design aspects. In light of the above, in designing a ship the naval architect tries to predict what the hull's natural frequencies of vertical vibration will be. He or she then selects a propeller rpm that will not coincide with any of those natural periods. Ideally, the design rpm will be somewhat below the lowest (that is, two-node) frequency of the hull. Attention must also be paid to the frequency of blade encounter with variations in the wake. This means that the product of propeller rpm and number of blades must also be kept away from those natural frequencies. The propeller itself should be dynamically, as well as statically, balanced before installation.

Fast naval vessels and many small craft feature one or more propellers situated under a flat, upward-sloping stern. In those cases few vibration problems will occur if the top of the propeller is kept at least a quarter of its diameter below the hull. In normal single-screw merchant ship hull forms, however, variable wake is always a threatening source of vibration. This is minimized by providing ample hull clearances both over the propeller and forward of it. A common rule of thumb is to allow a clearance of about 15 percent of the diameter above the propeller and 30 percent forward of it. Raked blades may be employed to help in this.

Propeller vibrations, when they occur, are most keenly felt in the aftermost 25 percent of the vessel's length. This is unfortunate because most modern merchant ships are arranged with the accommodations and working spaces near the stern. Thus the importance of vibration control is given added emphasis by this trend in design.

Engine-generated vibrations can be minimized by fitting resilient mountings between the engine and its foundation. The foundation itself should be of good strength and tied into specially strengthened hull members. In smaller units flexible shaft couplings may be employed.

Avoiding vibration in the crew accommodations is particularly important, so the naval architect will make a special effort to analyze the natural frequencies of the deckhouse. In some cases the entire deckhouse structure has been mounted on shock absorbing foundations. This has been effective in reducing the incidence of engine noise, but has proven less successful in eliminating propeller-induced vibrations.

Naval architects recognize that neither theory nor practice will ever eliminate vibrations 100 percent. This is because each little structural panel has its own natural frequencies of vibration, and is subject to a multiplicity of external exciting forces. (I was once told of a captain who complained that his bed springs continually vibrated. An exhaustive study pinned the blame on a small pump in the engine room several hundred feet away.) Given that some vibration is inevitable, the question then arises: how much is acceptable? To answer this question, charts have been devised showing combinations of frequency and amplitude of vibration that humans are likely to consider good, acceptable, or bad. For examples, see Vorus (1988) and Schellenberg (1980).

2.4 Post-trial modifications. Despite careful design-stage analysis, some degree of vibration may become evident when the ship is placed on trials. What steps may be employed to effect a cure? If the propeller is to blame, it may be replaced by one with a different number of blades, or perhaps it can be modified by trimming the blade tips. Another approach is to try to make the wake more uniform. This may be done by fitting a shelf-like appendage over the propeller. An alternative scheme is to fit a funnel-like appendage just forward of the upper part of the propeller. Figure 15.6 shows such a device.

If the trouble is found to arise with the main engine, the manufacturer may be able to effect modifications that will eliminate some of the exciting forces. Perhaps the foundation and associated hull structure should both be reinforced.

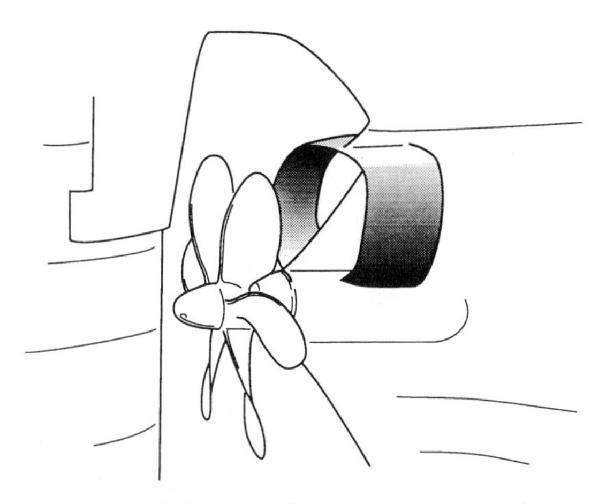

Fig. 15.6 Schneekluth wake-equalizing duct.

There will almost inevitably be found individual structural panels that will resonate with some item of machinery. These can normally be fixed by appending one or more stiffening members, which will change the natural frequencies of vibration and so avoid resonance.

2.5 Noise. The single best way to minimize noise problems is to arrange the vessel so as to separate passengers and crew from the source. Next, pay careful attention to what has been said above about placing machinery on flexible mountings or stout foundations. Especially noisy smaller items may be encased in soundproofed boxes. In small craft engines, wet exhausts are preferable to dry exhausts. The engine room itself can be sheathed in sound-absorbing materials on bulkheads and decks. Vent ducts and pipes should be fitted with flexible sleeves. Despite all these precautions, some noise will still find its way into the living and working spaces. The annoyance can be minimized through generous application of internal sound-absorbing materials: carpeted decks, drapes, and acoustical ceiling tiles.

Further Reading

Vorus, William S., "Vibration," Chapter 7 in *Principles of Naval Architecture,* Edward V. Lewis, Ed., Society of Naval Architects and Marine Engineers, Jersey City, N.J., 1988.

Hamlin, Cyrus, *Preliminary Design of Boats and Ships,* Cornell Maritime Press, Centreville, Md., 1989.

Schellenberg, Victor R., "Control of the Ship's Interior Environment," Chapter XIII in *Ship Design and Construction,* Robert Taggart, Ed., Society of Naval Architects and Marine Engineers, Jersey City, N.J., 1980.

3. Tonnage Measurement

3.1 Perspective. The first thing to be said here is that I am *not* speaking of weight-related units such as long tons or metric tons. Our subject, rather, concerns certain legally-defined measures of a ship's overall size or cargo hold volume, both measured in units of 100 cubic feet, or equivalent. These are called *registered tons,* and the subject is called *tonnage measurement* or perhaps *admeasurement.* The tonnage laws for the most part apply only to commercial ships whether intended for cargo or passengers. Shipowners and ship designers pay attention to the laws because many economic factors hinge on the tonnage of the individual ship. Taxes, harbor fees, and drydock fees are all based on tonnage. Many laws, such as those dealing with crew licensing, are stated in terms of tonnage. What all this means is that the shipowner will want the naval architect to design a ship having the smallest possible tonnage measurements while at the same time being big enough to do the work intended.

3.2 History. The subject is necessarily tied to history. As tonnage laws are changed, efforts are always made to protect owners against being suddenly placed at an economic disadvantage vis-à-vis their competitors. What was a relatively low-tonnage ship before the change should also be a relatively low-tonnage ship after the change. Therefore, to understand current tonnage laws we must look back into history.

In 15th century England, wine from the continent was a major import. To raise revenue the king applied an import duty on it and, to discourage smuggling, he decreed that the wine be carried in oversize (252-gallon) barrels called *tuns.* Harbor fees were accordingly based on a ship's cargo capacity as measured in those tuns, which were easily counted as the cargo was being discharged. But what if a ship came along carrying cargo packed in something other than barrels? Its equivalent "tunnage" capacity would then be derived from the product of its length, beam, and depth. Later, since depth was not always easy to measure, the authorities assumed it would equal half the beam, and so substituted $B/2$ for D in the formula. That led to tonnage measurements being proportional to the length times the beam squared. That was reasonable for the then existing ships, but shipowners quickly caught on to the fact that in any new ship, a deep hull coupled with a narrow beam would lead to lower fees and taxes. As you may at once guess, the eventual result was the loss of ships through lack of sufficient stability and consequent capsizings. Some reform was obviously in order.

At this point let me pause to remind you what was said in Chapter I, namely that each of those 252-gallon barrels full of wine happened to weigh 2240 lb, and that was the source of our long ton.

In 1854 an English naval architect named George Moorsom proposed a new approach to tonnage measurement that eliminated the shortcomings of the old system. His proposals, which were accepted worldwide, embraced two measurements: *gross tonnage,* a measure of the ship's overall size, and *net tonnage,* a measure of its cargo hold volume. Moreover, the values were to be derived from internal measurements in feet, to be converted to volumes by application of Simpson's rule. Measurements on typical ships showed that if the derived internal volume (in cubic feet) were divided by 100, the new tonnage measurement would be reasonably close to the old. And so the unit of tonnage measurement was changed from oversize barrels to units of 100 cubic feet.

Under the Moorsom system, certain hard-to-reach spaces, such as doublebottoms, were never measured. These were given the technical term *exemptions.* What remained to be measured was the gross tonnage. Then the volumes of the nonrevenue-producing spaces were subtracted, leading to the net tonnage. Although this system was sound in theory, as years went by, many loopholes were nibbled into the fabric of the law, some of them leading to unfortunate losses in seaworthiness. Moreover, each maritime nation had its own set of laws, and the nations with more rigorous rules inflicted an economic penalty on their shipowners. Clearly, reforms were once again in order.

3.3 Present law. In 1969, after long debate, a new approach to tonnage measurement arose from an international conference sponsored by IMCO (International Maritime Consultative Organization) of the United Nations. In 1982 the new rules became effective worldwide. Once again the gross tonnage is derived from the ship's entire internal volume, but now the measurements are usually taken off the drawings, so doublebottoms and so forth are easily included. Another change is that the measurements are in meters, but the final outcome yields numbers that are equivalent to hundreds of cubic feet. Thus the change in the law has had minimal effect on the tonnage values arising therefrom. Under the new law the net tonnage is derived from the actual volume of the cargo holds, with a correction factor that reduces the value in ships with generous freeboards. Thus, exemptions and deductions are things of the past. The new rules seem to be soundly thought out, and we may expect them to be found suitable for a good many years into the future.

I have omitted several of the less important details in what is said above. The main points are there, however, and should give you a fair idea of what registered tonnage measurements are all about.

3.4 Other tonnage rules. In addition to the international tonnage rules outlined above, the Panama and Suez Canals each has its own set of tonnage rules. These are used as a basis for fees charged to ships using those canals.

Further Reading

Cunningham, R. T. and Stitt, Phillips, "Tonnage Measurement," Chapter V in *Ship Design and Construction,* Robert Taggart, Ed., Society of Naval Architects and Marine Engineers, Jersey City, N.J., 1980.

4. Mooring

The subject of mooring has two main divisions. One has to do with securing a vessel alongside a fixed structure such as a pier or quay. The other has to do with anchoring the vessel in shallow water, usually allowing it to swing with wind and current. In either case the naval architect may want to predict the imposed loads likely to be encountered in service and to specify suitable gear. Few vessels are equipped to withstand infrequent extreme loads such as hurricanes. When such extremes are expected, the vessel is better off if taken out to the open sea. The various classification societies have rules that specify minimum requirements for mooring gear. These are based on past experience and in all but a few special cases will be found satisfactory for ordinary ships.

4.1 Mooring to fixed structures. When secured to a fixed structure, a vessel should be so attached that it can rise or fall with the tide, if any, and different conditions of loading. In many cases the arrangement should be such that the vessel may occasionally be moved along the structure so as to be repositioned under fixed shoreside cargo gear. The mooring system should be strong enough to withstand reasonable degrees of loading imposed by currents, tides, winds, and reactions from passing ships. The connection between ship and shore is effected by ropes (more correctly called *mooring lines*) extending between fixed fittings on the shore called *bollards* and some sort of hardware on the deck of the boat or ship. In small craft, single lines, bow and stern, are usually all that are needed. The lines are normally made of synthetic material and, being light in weight, can be handled manually. This sounds too obvious to be worth saying, but they should be of ample length and of good quality. I am continually impressed with the number of $50 000 speedboats I see tied up with short, ragged lengths of cheap rope.

In ships of any size, more sophisticated systems are required. Mechanical gear will be used to control the lines, which may be made of synthetic material or wire rope. If synthetic lines are employed, they will usually be pulled in (or slowed down when paying out) by being given a few turns around a powered drum of some sort. Once the ship is in the desired position, the lines will be taken off the drum and wrapped in figure eights around twin vertical posts called *bitts.* Each wire rope, on the other hand, will almost always be stored on a powered reel. When the shore end of the wire rope is attached to a fitting on the pier, the power can be applied to put tension in the line and pull the vessel closer to the pier. Once the ship is in position, the reel will automatically maintain the desired tension in the line. For example, if the tide is coming in, causing the ship to rise, the reel will automatically pay out increments of additional line so as to maintain the correct tension. With synthetic lines, on the other hand, such changes must be met by continual manual adjustment.

Figure 15.7 shows typical mooring arrangements for ships using synthetic lines and for ships using wire ropes.

Synthetic lines may be made of nylon, polyester, or polypropylene. Nylon is the strongest and most elastic. Polyester is intermediate in strength, least elastic, heaviest, and most resistant to abrasion. Polypropylene is least strong, but also least expensive, and lightest in weight. It is, however, subject to damage from prolonged exposure to sunlight (unless specially coated). The best quality synthetic lines are no longer formed of three twisted strands; instead, they are plaited or double-braided, either of which technique provides good flexibility without the traditional kinking problems found in the old triple-strand configuration.

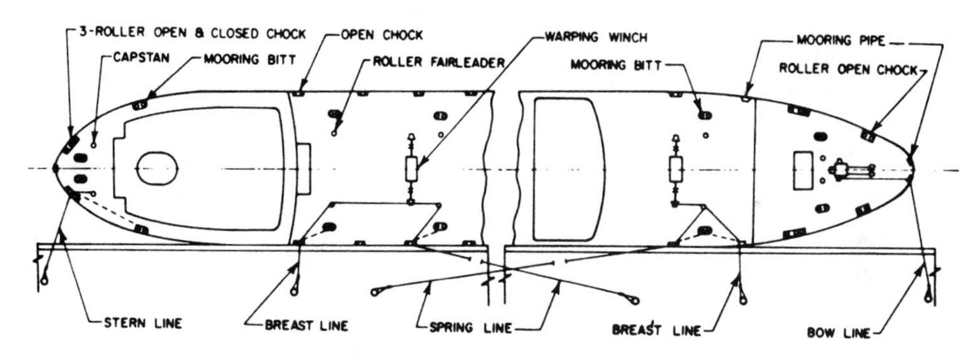

A. Conventional mooring arrangement with synthetic lines.

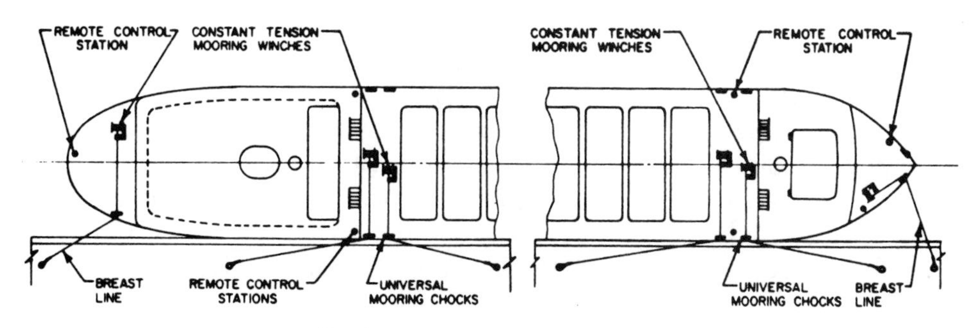

B. Arrangement with wire lines.

Fig. 15.7 Typical arrangements for mooring alongside a fixed structure (source: Hunley and Lemley, 1980).

When a vessel is approaching a pier, the mooring lines are fed from ship to shore by the traditional procedure of first tossing a light rope (or *heaving line*) to someone on the shore who will then use it to pull the mooring line up to a bollard. This operation will usually require several workers on the shore to wrestle each line into place. Once the line is looped over the bollard (the lines always have a big loop or *eye,* spliced into the shore end), the other end of the line is pulled taut by the mechanical equipment on deck. The procedure is repeated until the ship is securely tied up. Prudence dictates that at least four lines be employed: two at each end. Typically, however, at least six lines will be used (Fig. 15.7).

Special attention must be paid to the fittings where the mooring lines pass over the sides of the ship. These are called *chocks,* and they must be so designed and located as to allow easy passage of the lines in and out. Because wire ropes are not as flexible as their synthetic counterparts, their chocks must be specially shaped to avoid imposing harsh bends. This may be achieved with various combinations of grooved, free-turning wheels through which the wire is led. Combinations that allow freedom of motion in all directions are called *universal chocks.*

4.2 Mooring at anchor. The principal function of an anchor is to keep a vessel under passive control while in harbor waiting for a berth. The systems for doing this are essentially the same as those employed hundreds of years ago, which implies, correctly, that progress has been slow in this branch of maritime technology. Many refinements have been made and the scale is greatly enlarged, it is true. Still, our methods of anchoring a ship feature a good deal of brute strength and optimistic reliance on Poseidon's benevolence.

In small open boats the anchor may be a simple casting of lead or iron lowered on a length of rope, with dependence placed on the weight of the anchor to hold the boat in place. (True sons of the tillage may use a concrete building block.) In craft of somewhat larger size, the anchor will probably be of a sophisticated shape (in order to save weight); it will be attached with synthetic rope, and it will be pulled in by hand. As yet larger sizes are reached, mechanical help will be required for raising the anchor, and then chain will probably replace the rope. Most anchors work best when the tug of the rope or chain is applied nearly parallel to the bottom. This is achieved by using a length of rope or chain at least seven times the depth of the water.

As we advance to the case of large ships, we find that most carry at least three anchors. Two are carried in the bow and one in the stern. Usually all are designed so that they can be readily pulled up snug to the hull. Each is attached to a long, heavy chain that feeds into the anchor engine (the *windlass*) and then drops down into a deep storage bin called the *chain locker.* The extreme inboard end of the chain (appropriately called the *bitter end*) is attached to a reinforced fitting within the hull. In the typical anchoring situation, the ship relies on but one of its bower anchors. Two bower anchors, well spread out, may be used if yawing is found to be a problem. The stern anchor (or *stream anchor*) may also be used in concert with one or more of the bower anchors to prevent swinging. On occasion the stern anchor alone may be employed, principally in the case of a ship that finds it must anchor briefly while proceeding downstream.

In crowded harbors where little room is available for swinging at anchor, mooring buoys may be used instead of shipboard anchors. Each mooring buoy is held in permanent position by several anchors. The ship is attached to the buoy by means of one of its anchor chains or, weather permitting, by one of its mooring lines.

As mentioned earlier, when extreme weather is anticipated, the ship will probably be better off if taken out into the open sea. In intermediate conditions, if the ship is found to be dragging anchor, the engine may be started and the propeller turned just fast enough to take some of the load off the anchors.

Figure 15.8 shows three typical kinds of anchors. Type A, a stock anchor, reflects the form that reigned supreme a century ago. In the variation shown here, the horizontal cross bar (the *stock*) can be slid through a hole in the vertical part (the *shank*) and turned 90 degrees,

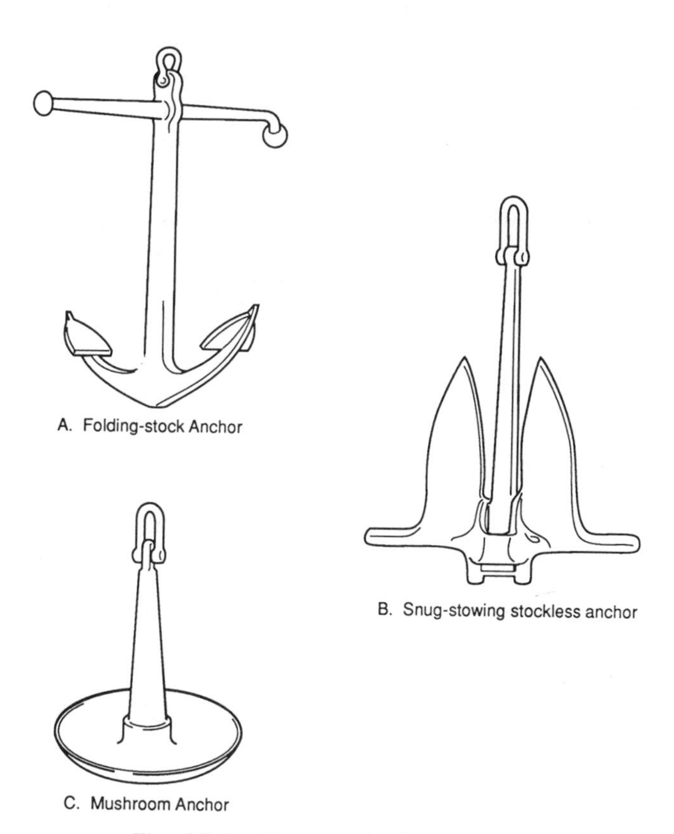

Fig. 15.8 Three typical anchors.

so that the anchor can be laid flat on deck. Type B is representative of the kind of anchor most commonly used on large ships. It has the virtue of being easy to stow. The shank is pulled up inside a tubular casting (the *hawsepipe*), leaving the spade-like appendages (the *flukes*) flat against the shell and out of the way. Type C is commonly used for permanent mooring buoys. It is relatively heavy, but has excellent holding characteristics.

Further Reading

Hunley, William H. and Lemley, Norman W., "Ship Maneuvering, Navigation and Motion Control," Chapter XII in *Ship Design and Construction,* Robert Taggart, Ed., Society of Naval Architects and Marine Engineers, Jersey City, N.J., 1980.

5. Aesthetics

How big a role does eye appeal play in naval architecture? There is no single answer. Looks are obviously important in yachts, sightseeing boats, and cruise ships. External appearance and interior decor merit careful thought, and attention is paid to changes in public taste. No one argues against investing heavily in making such vessels pleasant to view both inside and out. The attraction may arise through sheer beauty of form and color. Alternatively, it may result from eye-catching charm as evidenced in revivals of old-fashioned riverboat styles, with their paddle wheels, tall stacks, and generous helpings of gingerbread. In sailing yachts, beauty comes about almost automatically; but in other craft some price will usually have to be paid.

At the other end of the maritime scale, no one would argue in favor of spending money to beautify a garbage barge. But what about the vast majority of ships in between? Do the cargo ships of this world deserve an artistic touch? This is a matter of controversy. In recent decades the trend has been to ignore aesthetics in their design. Naval architects and owners argue that a merchant ship is an instrument of transport, not an item of floating scenery. A cargo ship's function is to increase wealth through socially useful work in moving goods from where they are found to where they are needed. In free-market economies, success in this is measured by profitability. Money spent in improving the ship's appearance will in no way improve its annual transport potential, nor will it reduce its operating costs. It can only subtract from its economic efficiency. Those who argue the other way (and I happen to be one) point out that there is more to economics than can be measured in dollars. The subject of economics is defined as the wise use of scarce resources. It is true that the main tool in economic analysis is the dollar sign. On the other hand, it is also true that ultimate success in any business venture gets down to human satisfaction, and many aspects of human satisfaction cannot be measured in monetary units. There is the matter of pride in owning, or crewing aboard, a good-looking ship. Could you not expect the crew members of a beautiful ship to exert some extra care in its maintenance? And if they gain some intangible satisfaction from serving aboard, are they not more likely to continue their seafaring careers?

Major improvements in appearance may be effected at only minor cost. A few carefully shaped curtain plates and fashion plates, and judicious selection of color schemes and patterns, can convert a box that only a philistine could love into an object of passing-fair appearance, and at less than a one percent increase in cost. An owner who is willing to make such an incremental investment may hope to attract and hold better crews, and he or she will certainly gain direct satisfaction from any model of the ship chosen to decorate home or office. Speaking in broader, more philosophical terms, does not the industry have a social contract to present to the world ships to please the eye? And if they honor that contract, are they not more likely to attract public support for political actions of benefit to the industry and to induce more bright young people to choose careers in the industry?

Quite aside from the matter of external appearance, there is that of interior decoration. Here everyone agrees that carefully chosen color schemes and lighting can do much to enhance shipboard living and working conditions. Bulkhead-covering patterns and textures are important. Psychologists have advice about selecting pictures to hang in public spaces: Reject marine scenes in favor of mountains, farms, or any other bucolic motif. Many naval architects recognize their need for help in all this and so employ interior decorators for advice in this phase of their work.

Further Reading

Barnaby, Kenneth C., *Basic Naval Architecture,* Hutchinsons, London, 1948.

Hamlin, Cyrus, *Preliminary Design of Boats and Ships,* Cornell Maritime Press, Centreville, Md., 1989.

Muller, W. H., "Some Notes on the Design of Crew Accommodations for Merchant Vessels," *Transactions,* Society of Naval Architects and Marine Engineers, Vol. 67, Jersey City, N.J., 1959, pp. 715–756.

6. The Venetian Gondola

I have chosen to put the frosting on this literary cake with a little essay on one of my favorite kinds of boats: the Venetian gondola.

These lovely craft embody all that is said above about aesthetics and also involve some thought-provoking mysteries of steering, propulsion, and hydrodynamics.

6.1 History. The Venetian gondola of today is a floating museum piece. Its design can be traced back with little change for hundreds of years. At one point the Doge of Venice became concerned that the wealthier citizens were wasting too much of the republic's wealth on ostentatious gondolas. This led him to decree that a standard design be adopted, with a plain black hull, save for the now-familiar silver-painted stem piece resembling a battle axe, and a modest scroll at the stern. Some all-but-invisible changes followed, but few, if any, have occurred since the 18th century.

At one time most gondolas were propelled by two or more men. As a matter of economy, however, it became increasingly common for the work to be done single-handedly. Traffic on the canals tends to be heavy, so the gondolier always faces forward so he can see ahead and avoid collisions. In some cases the gondolier may use two oars, but, because of the number of narrow canals, the usual gondolier gets by with but a single oar, which is always on the starboard side.

6.2 Overcoming asymmetry. The question then arises, with the thrust of the oar off to the starboard side, how does the gondolier keep his boat from swinging off to port? There are several parts to the answer. First, the hulls are asymmetric, being wider on the port side. (The apparent centerline is about six inches to port of the straight line running stem to stern; see Fig. 15.9.) When the gondola is operating in the design condition, that is, with passengers aboard, the extra buoyancy in the port side gives it a slight heel to starboard. The boat also has some trim aft. Because of the combined heel and trim, the flat bottom then becomes a gently sloping paravane, exerting a force tending to push the bow to starboard. Second, during the back stroke the gondolier leaves the oar in the water but gradually raises it so that it reaches the surface just before applying another power stroke. During this backstroke he feathers (twists) the blade of the oar so that the lateral pressure arising from the upward motion tends to swing the stern to port, and the bow to starboard. Third, the gondolier aligns the stem somewhat to starboard of his intended destination. This means that the gondola is sidling through the water, with the apparent course being off to starboard of the boat's actual direction of motion. This also tends to swing the bow to starboard.

Gondolas may be of ancient design, but they are far more attractive and sophisticated than many boats of modern vintage.

Further Reading

Pizzarello, Ugo and Pergolis, Riccardo, *The Boats of Venice,* translated by Coales and Pergolis, L'Altra Riva, Venice, 1981.

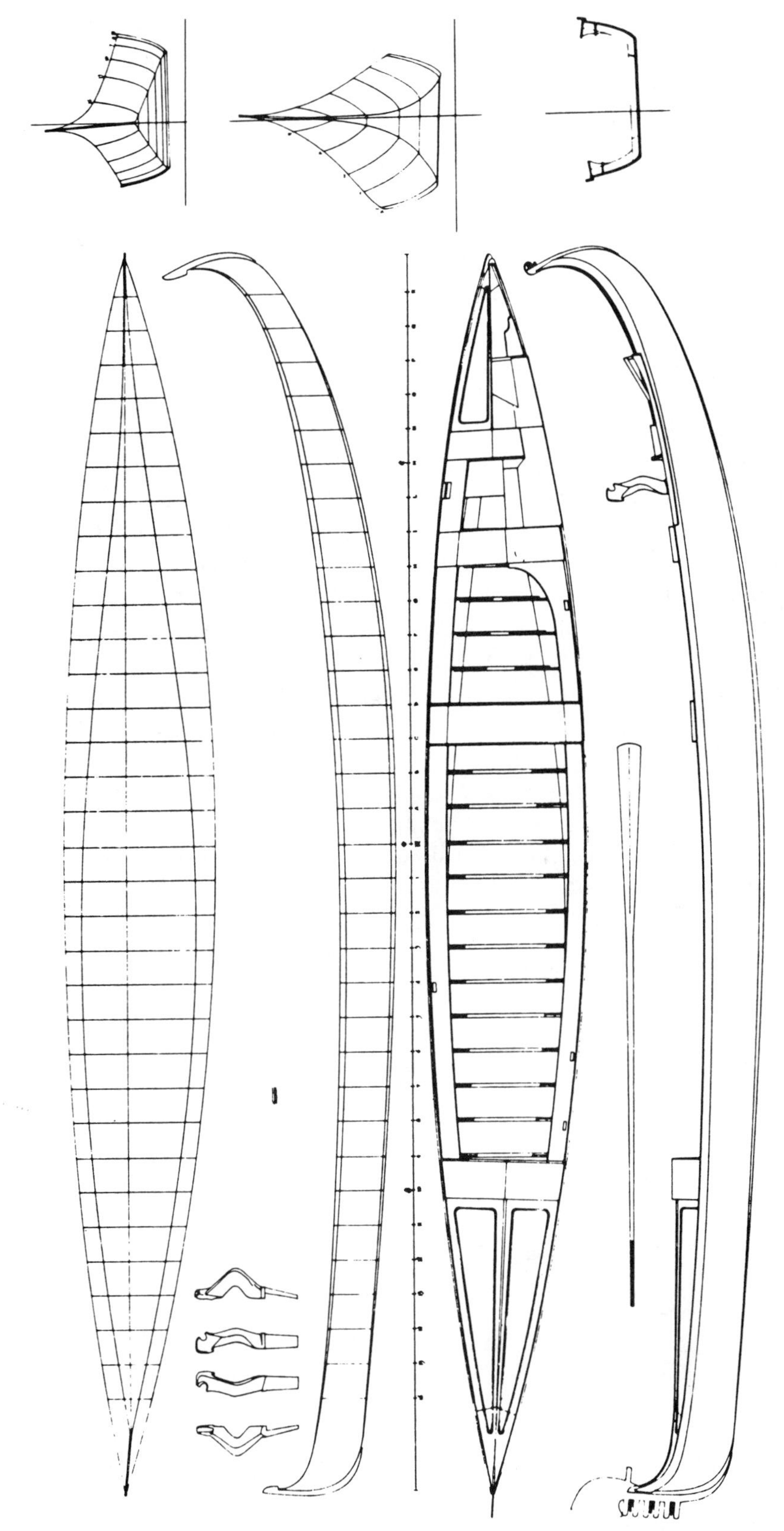

Fig. 15.9 Lines of a Venetian gondola (from Pizzarello and Pergolis, 1981).

Hamblin, Dora Jane, "Delicious! Ah! What Else Is Like a Gondola?" *Smithsonian,* Vol. 18, No. 4, July 1987, pp. 96–105.

7. Closure

Clearly, the subject of naval architecture is far broader and deeper than I have been able to spell out in this modest book. Nevertheless, I hope that what you have gained from these pages will serve as a reasonable introduction. More than that, I hope my literary efforts will not only give you a basic knowledge of the subject, but will also imbue you with some of the devotion that we naval architects feel toward our work. Taken as a whole, I believe that no other branch of engineering brings to its practitioners more pleasure and satisfaction than that enjoyed by naval architects. Welcome to the feast!

If you want to know more about any particular topic, the text contains numerous suggested references. In addition, for a comprehensive educational program, let me call to your attention the videographed series *Basic Naval Architecture* prepared by the National Shipbuilding Research Program (sponsored by the U.S. Navy and the Maritime Administration). For a catalog, write to Transportation Research Institute, University of Michigan, 2901 Baxter Rd., Ann Arbor, MI 48109-2150.

Finally, if by chance you are inspired to consider a career as a naval architect, what is your next step? Simple enough: Write to the Society of Naval Architects and Marine Engineers and ask for their career booklet. Their address is 601 Pavonia Ave., Jersey City, NJ 07306. Give it a thought!

Fig. 15.10 Aesthetics from the past. The recently built cruise ship *Pilgrim Belle* (now *Victorian Empress*) turns to an earlier period for exterior design inspiration. Although modern in every way in the internal arrangements, this ship has an external visual appeal that is hard to beat. Courtesy John W. Gilbert Associates, Inc.

Fig. 15.11 Lester Rosenblatt's ketch *Rosa II*. Good aesthetics come easily to a well-designed, well-handled sailing yacht. Courtesy M. Roseblatt and Son, Inc.

Glossary

Derived in part from *Ship Design and Construction,* Robert Taggart, Ed., SNAME, Jersey City, N.J., 1980.

Abaft. Aft of; toward the stern from a designated location.

Accommodation ladder. A portable, inclined ladder hinged to a platform attached to the edge of a ship, or at the sill of a shell entry port, and which can be positioned to provide access between ship and small craft or shore.

Adjustable-pitch propeller. A propeller in which the individual blades are fastened to the hub in such a way that they can, on occasion, be twisted, or removed and replaced.

Admeasurement. (See **registered tonnage.**)

Advance. The theoretical linear distance a screw propeller would move forward were it working within a nut. Also the forward distance traveled by a vessel after starting the hard-over turning maneuver. (Measured when heading has changed 90 deg.)

Aft. Toward, at, or near the stern.

Afterbody. That portion of a ship's hull abaft amidships.

Afterpeak. The compartment in the stern, abaft the aftermost watertight bulkhead.

After perpendicular. (See **length between perpendiculars.**)

Air-cushion vehicle. A vessel that uses powerful fans to inject air beneath itself, which provides vertical support allowing the craft to move with little friction over the surface.

Air port. A hinged glass window generally circular, in the ship's side or deckhouse, for light and ventilation; also called porthole, portlight, or side scuttle.

All-hatch ship. A vessel in which most of the deck over the cargo holds is cut away for hatches.

Alternator. (See **generator.**)

Amidships. In the vicinity of the mid-length of a ship as distinguished from the ends. Technically it is exactly half way between the forward and after perpendiculars. Also a helm order to indicate that the rudder is to be put on the centerline.

Anchor. A heavy forging or casting so shaped as to grip the sea bottom, and by means of a cable or rope, hold a ship or other floating structure in a desired position regardless of wind and current.

Anode. An electrode carrying a positive charge.

Apparent wind. The direction from which the air impinges on a boat in motion.

Appendages. The portions of a vessel extending beyond the main hull outline, including such items as rudder, shafting, struts, bossings, and bilge keels.

Arc welding. (See **electric arc welding.**)

Armature. The metallic skeleton of a ferrocement hull.

Athwartship. Across the ship, at right angles to the fore-and-aft centerline.

Automation. Automatic, rather than human, control systems.

Auxiliary machinery. All machinery other than that required for main propulsion.

Balanced rudder. A steering foil in which the turning stock is located close to the foil's center of lateral pressure.

Bale cubic. The cubic capacity of a cargo hold measured to the inside of the frames or cargo battens and underside of deck beams.

Ballast. Any solid or liquid weight placed in a ship to increase the draft, to change the trim, or to regulate the stability.

Ballast tank. A watertight enclosure that may be used to carry water ballast.

Bar keel. A longitudinal appendage, rectangular in section, running along the bottom of the ship, usually on the centerline.

Baseline. A fore-and-aft reference line. On large vessels it is at the upper surface of the flat plate keel at the centerline. Vertical dimensions are measured from a horizontal plane through the baseline, often called the **molded baseline.**

Battens. (See **cargo battens.**)

Beam, deck. An athwartship horizontal structural member, usually a rolled shape, supporting a deck or flat.

Beam knee. (See **knee, beam.**)

Beam, molded. The maximum breadth of the hull measured between the inboard surfaces of the side shell plating.

Beams, cant. Deck supports at aft end, arranged in a fan-like pattern.

Beam seas. A train of waves approaching a vessel directly from one side.

Beam, transom. (See **transom beam.**)

Bearding line. The intersection of the inside surface of the shell plating and the stem or sternpost.

Below. Indicative of a location below deck.

Bending moment. The product of a force acting at a distance from the support.

Bermuda sail. Same as jib-headed or Marconi sail.

Berth. Where a ship is docked or tied up; also a place to sleep aboard; a bunk or bed.

Between decks. (See **'tween decks.**)

Bevel. The angle between the flanges of a frame or other member (when greater than a right angle, open bevel; when less, closed or shut); also, to chamfer.

Bilge. Intersection of bottom and side. May be rounded or angular as in a chine-form hull. The lower parts of holds, tanks and machinery spaces where bilge water may accumulate.

Bilge and ballast system. A system of piping generally located in the holds or lower compartments of a ship and connected to pumps. This system is used for pumping overboard accumulations of water in holds and compartments and also for filling or emptying ballast tanks.

Bilge bracket. A vertical transverse flat plate welded to the tank top or margin plate and to the frame in the area of the bilge.

Bilge diagonal. An imaginary sloping plane whose location is shown on the body plan of a lines drawing, and whose true shape is shown, usually, below the half-breadth plan.

Bilge keel. A long longitudinal fin fitted at the turn of the bilge to reduce rolling.

Bilge strake. Course of shell plates at the bilge.

Bilge water. Stagnant water collected in the lower parts of a vessel.

Billboard. A sloping plate above the rake or gunwale of a vessel having low freeboard, on which an anchor is stowed and from which it can be released for use.

Binnacle. A stand or box for holding and illuminating a compass.

Bitter end. The inboard end of a ship's anchoring cable which is secured in the chain locker.

Bitt, mooring. Short metal column (usually two) extending up from a base plate attached to the deck for the purpose of securing and belaying ropes, hawsers, etc., used to secure a ship to a pier or tugboat. Also called a bollard.

Bitumastic. An elastic bituminous cement used in place of paint to protect steel.

Blade area ratio. That proportion of the propeller disk occupied by the blades.

Bleeder. A small cock, valve, or plug to drain off small quantities of fluids from a container or piping system.

Block coefficient. The ratio of the underwater volume of a ship to the volume of a rectangular block, the dimensions of which are the length, draft, and beam. The relationship is expressed as a decimal.

Body plan. A drawing consisting of two half-transverse elevations or end views of a ship, both having a common vertical centerline, so that the right-hand side represents the ship as seen from ahead and the left-hand side as seen from astern. On the body plan appear the forms of the various cross sections, the curvature of the deck lines at the side, and the projections, as straight lines of the waterlines, the buttock lines, and the diagonal lines.

Bollard. (See **bitt, mooring.**)

Bolster. (See **hawsepipe.**)

Bonjean curves. A set of curves, each of which represents a plot of the cumulative area of a station on the lines plan, from the baseline to any point above it.

Booby hatch. An access hatch in a weather deck, protected by a hood from sea and weather. Also called a **companionway.**

Boom. A long, round spar hinged at its lower end, usually to a mast, and supported by a wire rope or tackle from aloft to the upper end of the boom. Cargo, stores, etc., are lifted by tackle leading from the upper end of the boom. Also the spar at the bottom of a fore-and-aft sail.

Boom crutch. A term applied to a light structure built up from a deck to support the free end of a boom when it is not in use. Also called a boom rest.

Boom table. A stout, small platform usually attached to a mast to support the hinged heel bearings of booms and to provide proper working clearances when a number of booms are installed on or around one mast. Also called a **mast table.**

Boottop, or boottopping. The outer surface of the shell plating between light and load waterlines.

Bossing or boss. The curved swelling outboard portion of ship's shell plating that surrounds and supports the propeller shaft.

Boundary layer. The body of water that is dragged along with a vessel because of friction.

Bow. The forward end of a ship.

Bower anchor. An anchor carried at the bow.

Box girder. A large structural beam composed of four plates arranged in a rectangular configuration in cross section.

Bracket. A plate used to connect rigidly two or more structural parts, such as deck beam to frame, or bulkhead stiffener to the deck or tank top (usually triangular in shape).

Brake horsepower. The power delivered by a prime mover before entering any reduction gear.

Braking flaps. Hinged plates, usually attached to a rudder, that can be swung out so as to help stop a ship's forward motion.

Breadth, molded. (See **beam, molded.**)

Break. The end of a partial superstructure such as a poop, bridge, or forecastle where it drops to the deck below.

Breakwater. Inclined bulwark-like structure on a weather deck to deflect seawater coming over the bow and moving aft.

Breasthook. A triangular plate bracket joining port and starboard side stringers at the stem.

Breast line. A mooring rope oriented at about a right angle to the length of the ship.

Bridge. A superstructure at or near a ship's mid-length. (See also **bridge, navigating.**)

Bridge, flying. The platform forming the top of the pilothouse.

Bridge house. A term applied to an erection fitted on the upper or superstructure deck of a ship. The officers' quarters, staterooms, and accommodations are usually located in the bridge house and the pilothouse is located above it.

Bridge, navigating. The conning station or command post of a ship.

Brittle fracture. A tensile failure that acts with such abruptness that the failed material does not "neck down."

Broach. To be turned 90 degrees off course, usually owing to wave action from astern.

Broken stowage. The spaces between and around cargo packages, including dunnage, and spaces not usable because of structural interferences.

Brow. A watershed over an air port; also a small, inclined ramp to allow passage of people or trucks over a hatch coaming or bulkhead door sills, etc.

Buckler. A portable cover secured over the deck opening of the hawsepipes and the chain pipes to restrict the flow of water through the openings.

Building basin. A structure essentially similar to a graving dock, in which one or more ships or parts of ships may be built at one time; no launching operation is required, the ship being floated by flooding the basin.

Bulb angle. An L-shaped stiffening member with an enlargement running along the edge of the longer leg.

Bulb plate. A stiffening member resembling a flat bar, but with an enlargement running along one edge.

Bulk cargo. Cargo made up of commodities such as oil, coal, ore, grain, etc., and not shipped in bags or containers.

Bulkhead. A term applied to the vertical partition walls that divide the interior of a ship into compartments or rooms. The various types of bulkheads are distinguished by their location, use, kind of material, or method of fabrication, such as forepeak, longitudinal, transverse, watertight, wire mesh, pilaster, etc. Bulkheads which contribute to the strength of a vessel are called **strength bulkheads,** those which are essential to the watertight subdivision are **watertight** or **oiltight bulkheads,** and **gas-tight bulkheads** serve to prevent the passage of gas or fumes.

Bulkhead, afterpeak. A term applied to the first main transverse bulkhead forward of the sternpost. This bulkhead forms the forward boundary of the afterpeak tank.

Bulkhead, collision or forepeak. The foremost main transverse watertight bulkhead. It extends from the bottom shell to the freeboard deck and is designed to keep water out of the forward hold in case of bow collision damage.

Bulkhead deck. The uppermost deck up to which the transverse watertight bulkheads are carried.

Bulkhead, screen. A term applied to a light nonwatertight transverse bulkhead fitted in some Great Lakes bulk carriers. Its flexibility allows it to survive the impacts of the unloading machinery.

Bulkhead, swash. (See **swash bulkhead.**)

Bulwark. Fore-and-aft vertical plating immediately above the upper edge of the sheer strake.

Bunk. A berth or bed, usually built in.

Bunker. A place to store fuel. As a verb: to take on fuel.

Bunkers. Fuel.

Butt. The end joint between two plates or other members that meet end to end.

Buttock. The intersection of the molded surface of the hull with any vertical longitudinal plane not on the centerline.

Calk or caulk. To fill seams in a wood deck with oakum and pay them with pitch, marine glue, etc.

Camber. The rise or crown of a deck, athwartship; also called **round of beam.**

Camel. A fender to keep a vessel away from a pier or quay to prevent damage to the hull or pier; usually a floating body with massive padding of rope, tires, etc.

Cant frame. A frame not square to the centerline at the counter of the ship and connected at the upper end to the cant beams. (See **beams, cant.**)

Capacity plan. A plan outlining the spaces available for cargo, fuel, fresh water, water ballast, etc., and containing cubic or weight capacity lists for such spaces and a scale showing deadweight capacities at varying drafts and displacement.

Cap rail. Uppermost (usually flat) part of a railing.

Capstan. A warping head with a vertical axis used for handling mooring and other lines. It may have at its base a wildcat for handling anchor chain.

Cargo battens. Strips of wood fitted inside the frames to keep cargo away from hull structure; also called **sparring.**

Cargo port. Opening in a ship's side for loading and unloading cargo.

Casing, engine and boiler. Bulkheads enclosing large openings through decks above the engine and boiler rooms. This provides space for the boiler uptakes and access to these rooms, and permits installing or removing large propulsion units such as boilers or turbines.

Catamaran. A boat with twin, side-by-side hulls.

Cat boat. A sailboat with a single fore and aft sail.

Cathode. An electrode carrying a negative charge.

Cathodic protection. Protection of a ship's hull against corrosion by superimposing on the hull an impressed current provided by a remote power source through a small number of inert anodes. Also accomplished by fitting aluminum, magnesium, or zinc anodes in tanks or the underwater portion of a ship, which waste away by galvanic action.

Caught in stays. The condition when a sailing vessel, in trying to tack, stops turning just as it is headed right into the wind.

Cavitation. A phenomenon in which low pressure within a liquid allows vapor bubbles to form.

Ceiling, hold and tanktop. A covering usually of wood, placed over the tanktop for its protection.

Ceiling, joiner work. The overhead finished surface in quarters, etc.

Centerboard. A hinged plate that can be swung down through the slotted keel of a sailboat in order to enlarge the profile area and thus reduce leeway.

Center girder. A vertical plate on the ship's centerline between the flat keel and innerbottom or rider plate, extending the length of the ship. Also called **center vertical keel** (CVK) or **center keelson.**

Centerline. The middle line of the ship, extending from stem to stern at any level.

Chafing plate. Bent plate for minimizing chafing of ropes, as at hatches.

Chain locker. A compartment for the stowage of anchor chain.

Chain pipe. Pipe for passage of chain from windlass to chain locker.

Chain stopper. A device used to secure the chain cable when riding at anchor, thereby relieving the load on the windlass, and also for securing the anchor in the housed position in the hawsepipe. (See **devil's claw.**)

Chamfer. To cut off the sharp edge of a 90-deg corner. To trim to an acute angle.

Channel. A structural member comprising two flanges separated by a web attached to one edge of each flange.

Chevron frames. A hull-stiffening system employing a series of members arranged in a V-shaped configuration.

Chine. Abrupt change in transverse shape where a vessel's side and bottom come together.

Chock. A heavy smooth-surfaced fitting usually located near the edge of the weather deck through which wire ropes or fiber hawsers may be led, usually to piers. Also one of several pieces of metal precisely fitted between machinery units and their foundations to assure alignment, also made by pouring plastic material in place. Also a small piece of plate fitted to one side of a plated structure opposite the landing of a structural member on the other side.

Chock, boat. A cradle or support for a lifeboat.

Classification society. An organization that formulates rules for the construction of ships, monitors their construction and carries out inspections on ships in service to assure their continued seaworthiness.

Cleat. A fitting having two arms or horns around which ropes may be made fast.

Coaming, hatch. The vertical plating bounding a hatch for the purpose of stiffening the edges of the opening and resisting entry of water below.

Coast Guard. A unit of the federal government. Among its duties are those of assuring the seaworthiness of merchant ships and competence of the crews.

Cofferdam. Narrow void space between two bulkheads or floors that prevents leakage between the adjoining compartments.

Collision bulkhead. (See **bulkhead, collision.**)

Companionway. An access hatchway in a deck, with a ladder leading below, generally for the crew's use.

Compartmentation. The subdividing of the hull by transverse watertight bulkheads so that the ship may remain afloat under certain assumed conditions of flooding.

Containership. A vessel designed to carry cargo that is prepacked in large boxes in order to expedite loading and unloading.

Contraguide propeller. A twist introduced into the hull, or appendages, of a vessel in order to derive thrust from the spiraling energy of the water as it leaves the propeller.

Controllable-pitch propeller. A screw propeller in which the blades are pivoted and their angle modified by controls within the vessel.

Counter. (See **fantail.**)

Counterrotating propellers. Two screw propellers mounted on concentric shafts and turning in opposite directions.

Cowl. A hood-shaped top or end of a natural ventilation trunk that may be rotated in a direction to cause wind to blow air into or out of the trunk.

Crown. (See **camber.**)

Crow's nest. An elevated lookout station, usually attached to forward side of foremast.

Cryogenic. Pertaining to extremely low temperatures.

Cubic number. The product of the ship's length, beam, and depth divided by 100.

Curtain plate. A relatively narrow plate fitted at 90 degrees to the edge of an otherwise exposed deck plate, such as one finds alongside a deckhouse.

Curves of form. (See **hydrostatic curves.**)

Cycloidal propeller. A combined steering and propulsion device comprising a number of vertical blades arranged to rotate and revolve so as to give thrust in any desired direction.

Davit. A crane arm for handling lifeboats, anchors, stores, etc.

Day sailer. A sailboat intended for using over only short periods, none overnight.

Dead cover. A metal cover to close an air port from within the ship in case of breakage of the glass. Also called **deadlight.**

Dead flat. The portion of a ship's structure that has the same transverse shape as the midship frame. (See **parallel middlebody.**)

Deadlight or fixed light. A term applied to a portlight that does not open.

Deadrise. Athwartship rise of the bottom of the hull.

Deadweight. The carrying capacity of a ship at any draft and water density. Includes weight of cargo, dunnage, fuel, lubricating oil, fresh water in tanks, stores, passengers and baggage, crew and their effects.

Deadwood. A narrow part of the hull at the stern.

Deck. A platform in a ship corresponding to a floor in a building. It is the plating, planking, or covering of any tier of beams either in the hull or superstructure of a ship.

Deck beam. (See **beam, deck.**)

Deck, bulkhead. (See **bulkhead deck.**)

Deck, freeboard. Deck to which freeboard is measured; the uppermost continuous deck having permanent means of closing all weather openings.

Deck girder. A longitudinal web frame helping to support a deck.

Deck height. The vertical distance between the molded lines of two adjacent decks.

Deckhouse. An enclosed erection on or above the weather deck that does not extend from side to side of the ship.

Deck machinery. A term applied to steering gears, capstans, windlasses, winches, and miscellaneous machinery located on the decks of a ship.

Deck, platform. A lower deck, usually in the cargo space, which does not contribute to the longitudinal strength of the ship.

Deck, shelter. (See **shelter deck.**)

Deck stringer. The strake of deck plating that runs along the outboard edge of a deck.

Deck, weather. Uppermost continuous deck with no overhead protection.

Declivity. Inclination of shipways on which a ship slides during launching.

Deep tanks. Tanks extending from the bottom or innerbottom up to or higher than the lowest deck. They are often fitted with hatches so that they may also be used for dry cargo in lieu of fuel oil, ballast water, or liquid cargo.

Depth, molded. The vertical distance from the molded baseline to the top of the weather deck beam at side, measured at mid-length of the ship.

Derrick. A device for hoisting and lowering heavy weights, cargo, stores, etc.

Developable surface. A shape with curvature in a single direction to which flat materials can be forced without resort to extremes of temperature or pressure.

Devil's claw. A turnbuckle device having two heavy claws designed to fit over a link in the anchor chain for the purpose of securing the anchor chain.

Diesel engine. A form of internal combustion engine, which relies on the heat of compression to ignite the fuel mixture.

Dished. Pertaining to plating that is stiffened by being shaped in the form of one or more shallow troughs.

Displacement-length ratio. The number obtained by dividing a vessel's displacement by the (length/100) cubed. (Displacement in long tons, length in feet, as a rule.) It is indicative of the relative fatness of the underwater hull form. High values are indicative of short, fat forms.

Displacement, light. The weight of the ship complete including hull, machinery, outfit, equipment, and liquids in machinery.

Displacement, loaded. The displacement of a ship when floating at its greatest allowable draft. It is equal to the weight of water displaced and is the sum of the light displacement and the deadweight.

Displacement, total. The weight of water that would be displaced by the volume of the hull measured on the outer surface of the shell plating below the waterline. (See also **molded displacement.**)

Dock. A place for mooring a ship, usually between two piers.

Dog. A small metal fitting used to hold doors, hatch covers, manhole covers, etc., closed.

Doublebottom. Compartments at the bottom of a ship between innerbottom and the shell plating, used for ballast water, fresh water, fuel oil, etc.

Double chine. A transition between a vessel's side and bottom employing two longitudinal knuckles.

Double-skin. A structural arrangement featuring inner sides and innerbottom.

Doubling (doubler) plate. A plate fitted outside or inside of and faying (touching) against another to give extra local strength or stiffness.

Draft. The depth of the ship below the waterline measured vertically to the lowest part of the hull, propellers, or other reference point. When measured to the lowest projecting portion of the vessel, it is called the **extreme draft;** when measured at the bow, it is called **forward draft;** and when measured at the stern, the **after draft;** the average of the forward draft and the after draft is the **mean draft;** and the mean draft when in full load condition is the **load draft.** Also, in cargo handling, the unit of cargo being hoisted on or off the ship by the cargo gear at one particular hoist.

Draft marks. The numbers on each side of a ship at the bow and stern, and sometimes amidships, to indicate the distance from the lower edge of the number to the bottom of the keel or other fixed reference point. The numbers are 6 in. high and spaced 12 in. bottom to bottom vertically.

Drag. The designed excess of draft aft over that forward when fore and aft drafts are measured from the designed waterline.

Drift angle. The angle between a vessel's apparent and true courses.

Dry-bulk cargo. Commodities, other than in liquid form, that are carried aboard ship without benefit of packaging. Coal, grain, and various ores are examples.

Dry dock. A facility used in effecting repairs to the bottom of ships. There are two major kinds. A **floating drydock** is a hollow steel structure that can be submerged with water ballast. The ship is then positioned over the submerged dock. The ballast is then pumped out and the dock, on rising, lifts the ship out of the water. The other kind of repair facility is the **graving dock,** which is defined elsewhere in this Glossary.

Dunnage. Cushioning, loose material placed under or among cargo in the holds to prevent cargo motion or chafing.

Dutchman. A piece of steel fitted or driven into an opening to cover up open joints or crevices usually caused by poor workmanship.

Dynamic positioning. A means of holding a ship in a relatively fixed position with respect to the ocean floor without using anchors, accomplished by two or more propulsive devices controlled by inputs from sonic instruments on the sea bottom and on the ship, by gyrocompass, by satellite navigation, or other means.

Dynamic stability. The ability of a body to remain upright when subjected to various external disturbing influences.

Effective length. A length sometimes used for speed-power calculations and the coefficients therefor. It is determined from the sectional area curve by excluding any abrupt tailing off at the after end of the curve such as often occurs with single-screw, cruiser-stern ships. In multiscrew normal vessels, it is usually the load waterline length, but in single-screw ships with either cruiser or fantail sterns, it is usually the length from the forward perpendicular to about the middle of the propeller aperture.

Elastic limit. (See **proportional limit.**)

Electric arc welding. The common way of forming a joint between two pieces of metal. The process involves passing an electric current through an expendable metal rod held a short distance from the members to be joined. The electric arc thus generated melts the rod and the near surfaces of the metal parts so that all fuse into one.

Electrolysis. The effect of an electric current passing through an electrolyte in transferring ions from one part to another.

Engine casing. A vertical air shaft rising from over the engine room. It makes room for the exhaust system, ventilation ducts, piping, wiring, and access ladders.

Ensign staff. A flagstaff at the stern.

Entrance. That portion of a ship's underwater body forward of the parallel middlebody or the point at which the slope of the sectional area curve is zero.

Epoxy resin. Any of a group of materials derived by polymerization of certain epoxy chemicals.

Erect. To hoist into place and bolt up on the ways fabricated parts of a ship's hull, preparatory to fitting and welding.

Erection. A generic term for deckhouses and superstructures.

Ergonomics. The study of humans and machines working together.

Escape trunk. A vertical trunk fitted with a ladder to permit personnel to escape being trapped. Usually provided from the after end of the shaft tunnel to topside spaces.

Even keel. A ship is said to be on an even keel when the keel is horizontal.

Expanded metal. A metallic mesh made by slotting and then stretching a sheet of the material.

Expansion trunk or tank. A trunk extending above a space used for the stowage of liquid cargo. The surface of the cargo liquid is kept sufficiently high in the trunk to permit expansion without risk of excessive stress on the hull or of overflowing, and to allow contraction of the liquid without increase of free surface.

Extrusion. A metallic part whose sectional shape is formed by the metal being squeezed (while heated to plastic condition) through a hole of the desired shape.

Fabricate. To process hull material in the shops prior to assembly or erection. In hull work, fabrication consists of shearing, shaping, scarfing, rabbeting, and beveling.

Face plate. Generally a narrow stiffening plate fitted along the edge of web frames, stringers, etc., to form the flange of the member.

Fair. To smooth or fair up a ship's lines; eliminating irregularities; also to assemble the parts of a ship so that they will be fair, that is, without kinks, bumps, or waves.

Fairlead or fairleader. A fitting or device used to preserve or to change the direction of a rope so that it will be delivered on a straight lead to a sheave or drum.

Fairwater. A term applied to plating fitted around the ends of shaft tubes and strut barrels, and shaped to streamline the parts, thus eliminating abrupt changes in the water flow. Also applied to any casting or plating fitted to the hull for the purpose of preserving a smooth flow of water.

Fall. The rope used with blocks to make up a tackle. The end secured to the block is called the **standing part;** the opposite end, the **hauling part.**

Fantail. The overhanging stern section of ships with round or elliptical after endings to uppermost decks and which extend well abaft the after perpendicular. Also called **counter.** At times, a synonym for the stern.

Fashion plate. A rounded transition plate at the end of a bulwark or superstructure.

Fathom. A measure of length, equivalent to 6 linear feet, used for depths of water and lengths of anchor chain. (See **shot.**)

Fathometer. A device to measure the depth of water, by timing the travel of a sound wave from the ship to the ocean bottom and return.

Faying surface. The overlapping faces between two adjoining parts.

Fender. The term applied to devices built into or hung over the sides of a vessel to prevent the shell plating from rubbing or chafing against other ships or piers; a permanent hardwood or steel structure which runs fore and aft on the outside above the waterline and is firmly secured to the hull; wood spars, bundles of rope, automobile tires, woven cane, or covered cork hung over the sides by line when permanent fenders are not fitted.

Ferrocement. A structural material comprising a relatively thin layer of concrete intimately reinforced with steel rods and mesh.

Fiber glass. A structural material comprising glass fibers bonded by some form of plastic material. Also spelled **fiberglass.**

Fidley. The top of engine and boiler room casings on the weather deck. A partially raised deck over the engine and boiler casings, usually around the smokestack.

Filler metal. That part of a welding rod that melts and is deposited in a welded joint.

Fillet weld. A weld joining two structural members that come together at, or close to, a right angle.

Finger pier. A narrow pier. (See **pier.**)

Fitter. A shipyard worker skilled in arranging structural parts ready for welding.

Fixed light or deadlight. Circular non-opening window with glass in side of ship, door, skylight cover, etc.

Fixed-pitch propeller. A screw propeller made in one piece.

Flange. The part of a plate or shape bent at right angles to the main part; also to bend over to form an angle.

Flank speed. In naval vessels: top speed.

Flare. The spreading out of the hull form from the central vertical plane, with increasing rapidity as it rises from the waterline to the rail; usually in the forebody. Also a night distress signal.

Flat. A partial deck, usually without camber or sheer.

Flat bar. A structural component of simple rectangular shape in section.

Flexplate. A plate, at one end of a foundation, which permits a piece of machinery mounted on the foundation to expand or contract freely.

Floodable length. The length of ship which may be flooded without sinking below its safety or margin line. The floodable length of a vessel varies from point to point throughout its length and is usually greatest amidships and least near the quarter lengths.

Floor. Vertical transverse plate immediately above the bottom shell, often located at every frame, extending from bilge to bilge.

Floor, open. Within a doublebottom, a transverse framing member of skeleton configuration.

Floor, solid. Within a doublebottom, a transverse framing member comprising a stiffened vertical plate with lightening/access holes.

Flukes. The parts of an anchor that are intended to dig into the bottom.

Flush-deck ship. A ship constructed with an upper deck extending throughout its entire length without a break and without any superstructures, such as forecastle, bridge, or poop.

Fluted. Pertaining to plating that is stiffened by being shaped in a series of knuckles.

Fo'c's'le. The standard abbreviation, and way of saying, **forecastle.**

Fore. A term used indicating portions, or that part, of a ship at (or adjacent to) the bow. Also applied to that portion and parts of the ship lying between amidships and the stem; as **forebody, forehold,** and **foremast.**

Fore-and-aft. In line with the length of the ship; longitudinal.

Fore-and-aft sail. A sail hinged at its forward end.

Forebody. That portion of the ship's body forward of amidships.

Forecastle. A superstructure fitted at the extreme forward end of the upper deck.

Forefoot. The lower end of a ship's stem where it curves to meet the keel.

Forepeak. The watertight compartment at the extreme forward end. The forward trimming tank.

Forward. In the direction of the stem.

Forward or fore perpendicular. (See **length between perpendiculars.**)

Foundation. The structural supports for the boilers, main engines or turbines and reduction gears are called **main foundations.** Supports for machinery space auxiliary machinery are called **auxiliary foundations.** Deck machinery supports are called, for example, **steering engine foundation, winch foundation,** etc.

Founder. Sink and go to the bottom.

Four-stroke engine. In reciprocating prime movers, one in which energy is applied every other complete rotation.

Frame. A term used to designate one of the transverse members that make up the rib-like part of the skeleton of a ship. The frames act as stiffeners, holding the outside plating in shape and maintaining the transverse form of the ship. (See also **longitudinals.**)

Frame spacing. The fore-and-aft distance, heel to heel, of adjacent transverse frames.

Framing. A general term for the stiffening members in any part of the hull structure.

Freeboard. The distance from the waterline to the upper surface of the freeboard deck at side.

Freeboard deck. (See **deck, freeboard.**)

Freeing port. An opening in the lower portion of a bulwark that allows deck water to drain overboard. Some freeing ports have hinged gates that allow water to escape, but which swing shut to prevent seawater flowing inboard.

Free surface. The condition of liquid without any constraints on its upper level.

Frictional resistance. Inhibition to forward motion arising from viscosity between water and hull.

Froude number. A nondimensional number indicating the relation between a vessel's length and its speed. See Section 2.4 of Chapter V.

Full scantling ship. A ship designed with scantlings and weather deck closing arrangements qualifying the ship for minimum freeboard, measured from the uppermost continuous deck, according to the International Load Line Convention.

Funnel. (See **smokestack.**)

Furnaced plate. A plate that requires heating in order to be shaped.

Gaff. A sloping spar at the top of a fore-and-aft sail.

Gaff sail. A sail fitted with a gaff.

Galley. A cook room or kitchen on a ship.

Gallows. An approximately horseshoe-shaped structure used to raise nets or paravanes on a fishing vessel.

Gangway. A passageway, side shell opening, or ladderway used for boarding or leaving a ship.

Gantry crane. A hoisting device, usually traveling on rails, having the lifting hook suspended from a car which is movable horizontally in a direction transverse to the rails.

Garboard strake. The line of bottom shell plating adjacent to the keel.

Gasket. Flexible material used to pack joints in machinery, piping, doors, hatches, etc., to prevent leakage of liquids or gases.

Gas turbine. A prime mover that derives energy from internally burned fuel, the exhaust gases of which act against many small blades to turn the shaft.

General cargo. Goods to be transported in a mixture of forms, but usually packaged in some way.

Generator. A device used to convert mechanical energy into electrical energy.

Girder. A continuous member running fore and aft under a deck for the purpose of supporting the deck beams and deck. The girder is generally supported by widely spaced pillars. Also, the vertical fore-and-aft plate members in the bottom of single- or doublebottom ships.

Girth. Any expanded length, such as the length of a frame from gunwale to gunwale.

Gooseneck, or Pacific iron. A swivel fitting on the end of a boom for connecting it to the mast or mast table. It permits the boom to rotate laterally and to be peaked to any angle Also, a ventilation terminal in the weather deck, of rectangular cross section and consisting of a 180-degree bend with the opening facing down; usually fitted with screen and hinged cover.

Grain cubic. The volumetric capacity of a cargo hold measured to shell plating and underside of deck plating.

Graving dock. A structure for taking a ship out of water, consisting of an excavation in the shoreline to a depth at least equal to the draft of ships to be handled, closed at the water end by a movable gate, and provided with large-capacity pumps for removing water; blocks support the ship when the water is pumped out.

Grimm propeller. A variation on the counter-rotating propeller in which the trailing unit derives its energy from the spiraling water coming off the forward propeller and imparts it to extended blade tips of opposite pitch.

Grommet. A soft ring used under a nut or bolt head to maintain watertightness; also an eye fitted into canvas.

Gross tonnage. (See **tonnage, gross.**)

Ground-level building site. A facility wherein the ship is built without sloping building ways, the baseline being parallel to the water surface; means of physically moving the completed ship, or large components, to the water are required.

Ground tackle. A general term for anchors, cables, wire ropes, etc., used in anchoring a ship to the bottom.

Ground ways. The tracks on which a vessel slides when being launched.

Gudgeon. Bosses or lugs on sternpost drilled for the pins (pintles) on which the rudder hinges.

Gusset plate. A bracket plate lying in a horizontal or nearly horizontal plane.

Guy. A line used to brace a spar.

Gypsy head. A cylinder-like fitting on the end of winch or windlass shafts. Fiber line is hauled or slacked by winding a few turns around it, the free end being held taut manually as it rotates.

Half beam. A dimension defining half the vessel's width at the widest point.

Half-breadth plan. A part of the lines drawing in which are shown the shapes of the waterplanes. Also called the waterlines plan.

Half siding. Half of the width of the flat area of the keel.

Halliard, halyard. Lines used in hoisting sails, signals, flags, etc.

Hard spot. The intersection of two structural members where one is far stronger than the other.

Hatch (hatchway). An opening in a deck through which cargo and stores are loaded or unloaded.

Hatch, booby. (See **booby hatch**.)

Hatch coaming. (See **coaming, hatch**.)

Hawsepipe. Tube through which anchor chain is led overboard from the windlass wildcat on deck through the ship's side. Bolsters form rounded endings at the deck and shell to avoid sharp edges. Stockless anchors are usually stowed in the hawsepipe.

Head. Toilet; believed to be derived from "ship's head," when a small platform outside the bulwarks near the bow was the only semblance of sanitary facilities.

Headlog. In river craft of rectangular shape, the horizontal member at the extreme end between the rake shell plating and the deck. Usually a vertical plate of considerable thickness owing to its susceptibility to damage in service.

Headreach. In a crash stop, the forward transfer of distance after start of the stopping maneuver.

Heat exchanger. Any item of equipment that serves the purpose of using one fluid to heat or cool another. A radiator is a common example.

Heat-treatable aluminum alloys. Collectively, that family of aluminum alloys that depend on some form of heat treatment to develop their strength.

Heaving. Vertical translational motion of a vessel.

Heaving line. A light rope that, when tossed ashore, allows shore personnel to pull a mooring line from the ship to a mooring bitt.

Heel. The inclination of a ship to one side. (See **list.**) Also the corner of an angle, bulb angle, or channel, commonly used in reference to the molded line.

High-performance craft. Vessels (usually small) of exceptional performance capability in such factors as high speed.

Hogging. Straining of the ship that tends to make the bow and stern lower than the middle portion. (See **sagging.**)

Holds. Spaces below deck for the stowage of cargo; the lowermost cargo compartments.

Horn, rudder. A heavy casting or weldment projecting down from the hull immediately abaft the propeller, to support the gudgeon fitted to take the single pintle of a semibalanced rudder.

House, deck. (See **deckhouse.**)

Hull. The structural body of a ship, including shell plating, framing, deck, bulkheads, etc.

Hull efficiency. A concept that combines wake gain and thrust deduction as a step in relating effective horsepower and shaft horsepower.

Hull engineering. That collection of mechanical devices outside the engine room.

Hull girder. That part of the hull structural material effective in the longitudinal strength of the ship as a whole, which may be treated as analogous to a girder.
Hydrodynamics. The study of the physics of water in motion.
Hydrofoil boat. A small craft that, when driven at speed, has its hull lifted out of the water by the support of small wing-like surfaces in the water.
Hydroplane. A boat designed to obtain lift from the impact of water as the boat moves at speed along the surface.
Hydrostatic curves. Lines drawn on grid paper showing hull form characteristics (such as displacement and block coefficient) at different drafts. Also called **curves of form.**
I-beam. A rolled metal shape resembling the letter "I" in cross section.
Inboard. Inside the ship; toward the centerline.
Inboard profile. A drawing that shows the interior arrangements, at least in part, of a ship. It is what you would see if a giant saw were to cut the vessel along the centerline from end to end and then remove the near side.
Inclining experiment. A procedure for determining a vessel's metacentric height.
Innerbottom. Plating forming the top of the doublebottom; also called **tank top.**
Insert plate. A relatively thick plate inserted and welded into plating of normal thickness in order to provide added strength in highly loaded locations.
Intangible factors. Considerations in decision making that cannot readily be weighed in monetary units.
Intercostal. Made in separate parts: between floors, frames or beams, etc.; the opposite of continuous.
Inverted angle. A stiffening member shaped like the letter L, with the longer leg welded to a plate.
Isherwood system. A former proprietary system for longitudinal framing.
Jack staff. A flagstaff at the bow.
Jacob's ladder. A portable ladder with flexible or articulated sides used for access between deck and small craft or piers.
Jib. A triangular fore-and-aft sail fitted forward of the forward mast.
Jib-headed sail. A fore-and-aft sail of approximately triangular shape.
Joint efficiency. The ultimate strength of a fastening divided by the ultimate strength of the members joined.
Keel. The principal fore-and-aft component of a ship's framing, located along the centerline of the bottom and connected to the stem and stern frames. Floors or bottom transverses are attached to the keel.
Keel, bilge. (See **bilge keel.**)
Keel blocks. Heavy wood or concrete blocks on which ship rests during construction or dry docking.
Keel, center vertical. The vertical, centerline web of the keel (CVK).
Keel draft. Vertical distance between the waterline and the bottom of the keel.
Keel drag. (See **trim.**)
Keel, flat plate. The horizontal, centerline, bottom shell strake constituting the lower flange of the keel.
Keel rider. A plate running along the top of the center vertical keel in a single-bottom vessel.
Keelson, side. Fore-and-aft vertical plate member located above the bottom shell on each side of the center vertical keel and some distance therefrom.
Kerf. The material removed by a cutting device, such as a burning torch, in preparing a structural member.
Ketch. A boat with fore-and-aft sailing rig with two masts, the forward one being somewhat larger than the other.
Kingposts. Strong vertical posts used in pairs instead of a mast to support booms and rigging to form a derrick: also called **samson posts.**

Knee, beam. Bracket connecting a deck beam and frame.

Knight's head bow. A hinged arrangement of the structure at the forward end of a ship allowing it to be raised, usually as a means of access for wheeled vehicles.

Knot. A unit of speed, equaling one nautical mile per hour; the international nautical mile is 1852 m (6076.1 ft.).

Knuckle. An abrupt change in direction of the plating, frames, keel, deck, or other structure of a ship.

Kort nozzle. A circular shroud fitted around a screw propeller to enhance thrust.

Ladder, accommodation. (See **accommodation ladder.**)

Ladder, Jacob's. (See **Jacob's ladder.**)

Lap. A joint in which one part overlaps the other.

Lateen rig. An arrangement of fore and aft sails in which spars at top and bottom of each sail lie alongside the mast.

Launching ways. (See **ground ways** and **sliding ways.**)

Laying off. The development of the lines of ship's form on the mold loft floor and making templates therefrom; also called **laying down.**

Lee. The side away from the wind.

Lee boards. Hinged plates that can be swung down a sailboat's sides in order to reduce leeway.

Lee helm. A condition in which a sailboat can be kept on course only by turning the tiller somewhat downwind.

Leeward. Pertaining to the lee.

Leeway. The angle between a boat's apparent heading and its actual direction of motion. Also called **drift angle.**

Left-hand propeller. A screw propeller that provides forward thrust when turning counterclockwise as viewed from astern.

Length between perpendiculars. The length of a ship between the forward and after perpendiculars. The forward perpendicular is a vertical line at the intersection of the fore side of the stem and the summer load waterline. The after perpendicular is a vertical line at the intersection of the summer load line and the after side of the rudder post or sternpost, or the centerline of the rudderstock if there is no rudder post or sternpost.

Length, effective. (See **effective length.**)

Length overall. The extreme length of a ship measured from the foremost point of the stem to the aftermost part of the stern.

Lifeboat. A boat carried by a ship for use in emergency.

Life raft. A buoyant device, usually of inflatable material, designed to hold people abandoning ship.

Lightening hole. A hole cut in a structural member to reduce its weight.

Light ship weight. (See **displacement, light.**)

Limber hole. A small hole or slot in a frame or plate for the purpose of preventing water or oil from collecting; a drain hole.

Liner. A flat or tapered strip placed under a plate or shape to bring it into line with another part that it overlaps; a filler. Also, a commercial ship that sails between fixed ports on a fixed schedule.

Lines (plan). The drawing that shows the shape or form of the ship. (See also **molded lines.**)

Line shafting. Sections of main propulsion shafting between the machinery and the tailshaft.

Liquefied-gas carrier. A vessel specially designed to carry gas that has been liquefied by being compressed or cooled.

List. If the centerline plane of a ship is not vertical, as when there is more weight on one side than on the other, it is said to list, or to heel.

Liverpool head. A metal device fitted to the top of a smoke pipe, as from a galley, to minimize inflow of water and to promote the draft in the pipe. Sometimes mechanically activated by wind.

Load line. The deepest draft to which a vessel is allowed to load.

Loadline mark. Inscription on the side of a ship showing the maximum draft to which it may be loaded. Commonly called the **Plimsoll mark.**

Load on top. Oil separated from water (as the residue from cleaning cargo tanks) is loaded on top of the vessel's next cargo, rather than being pumped overboard without separation. Slop tanks are provided in which the separation takes place.

Load waterline. The line on the lines plan of a ship, representing the intersection of the ship's form with the plane of the water surface when the ship is floating at the summer freeboard draft or at the designed draft.

Lock. A civil engineering structure with gates at either end, used to move a vessel from one water level to another.

Lofting. The process of developing the size and shape of components of the ship from the designed lines; traditionally, making templates using full scale lines laid down on the floor of the mold loft; today, largely performed at small scale using photographic or computer methods.

Longitudinal center of flotation. The fore-and-aft location of the center of area of the ship's waterplane.

Longitudinal framing. A hull-stiffening system in which the stiffening members run along the length of the ship.

Longitudinal loading. The situation in which there is a difference between the distribution of the weights along the length of a ship and the distribution of the buoyant support.

Longitudinal prismatic coefficient. (See **prismatic coefficient.**)

Longitudinals. Fore-and-aft structural members attached to the underside of decks, flats, or to the innerbottom, or on the inboard side of the shell plating, in association with widely spaced transverses, in the longitudinal framing system.

Longitudinal stability. That branch of the study of stability having to do with questions of trim.

Long-tail boat. Small craft propelled by a deck-mounted gasoline engine directly connected to an extended propeller shaft.

Long tons. A unit of weight equal to 2240 lb.

Louver. An opening to the weather in a ventilation system, fitted with a series of overlapping vanes at about 45 degrees intended to minimize the admission of rain or spray to the opening. Also any opening fitted with sloping vanes.

Magnetohydrodynamic propulsion. A still-developing technique for producing thrust by using electromagnetic force to move seawater through a longitudinal duct.

Manhole. A round or oval hole cut in decks, tanks, etc., for the purpose of providing access.

Marconi sail. A fore-and-aft sail of approximately triangular shape. Same as jib-headed sail.

Margin bracket. A bracket connecting a side frame to the margin plate at the bilge; sometimes called bilge bracket.

Margin line. A line, not less than 3 in. below the top of the bulkhead deck at side, defining the highest permissible location on the side of the ship of any damaged waterplane in the final condition of sinkage, trim and heel.

Margin plate. The outboard strake of the innerbottom. When the margin plate is turned down at the bilge it forms the outboard boundary of the doublebottom, connecting the innerbottom to the shell plating at the bilge.

Marine engineer. A person with specific expertise in designing or selecting machinery for boats and ships and in making decisions on their arrangement within the hull. Also a ship's officer involved in operating and maintaining machinery.

Marine railway. An arrangement of tracks from shore to a sufficient depth of water to permit a vessel to be placed on a moveable carriage, which may then be drawn up the track in order to give access to the portion of the vessel below its waterline. Generally limited to comparatively small vessels.

Mast. A tall vertical or raked structure, usually of circular section, located on the centerline of a ship and used to carry navigation lights, radio antennae, and, often, cargo booms. (See **kingpost.**) In a sailing vessel it supports the sail(s) and associated spars.

Mast step. A term applied to the foundation on which a mast is erected.

Mast table. (See **boom table.**)

Maximum section coefficient, C_X. The ratio of the area of the maximum vertical transverse cross section of the underwater body of a ship to the product of the waterline beam and the draft at that section.

Mechanical door. A door that can be opened or closed by machinery, often with remote control.

Mensuration. The branch of geometry dealing with methods of measuring plane or three-dimensional, figures.

Merchant ship. A vessel intended for commercial transport.

Messroom. Dining room for officers or crew.

Metacenter. The center of buoyancy of a listed ship is not on the vertical centerline plane. The intersection of a vertical line drawn through the center of buoyancy of a slightly listed vessel intersects the centerline plane at a point called the metacenter.

Metacentric height. The distance from the metacenter to the center of gravity of a ship. If the center of gravity is below the metacenter, the vessel is initially stable.

Metacentric radius. The vertical separation of the center of buoyancy and the metacenter.

Midship. (See **amidships.**)

Midship coefficient. The ratio of the area of the vertical transverse cross section of the underwater body of a ship at mid-length divided by the product of the waterline beam and draft at that point.

Midship half-length. Half the length of a ship with its center at the ship's mid-length.

Midship section. A drawing showing a typical cross section of the hull and superstructure at or near amidships and giving the scantlings of the principal structural members.

Mizzen. Pertaining to the aftermost mast on a sailboat.

Mock-up. A three-dimensional, full-size replica of the shape of a portion of a vessel, used where the geometry makes fabrication of steel members from conventional templates difficult or to avoid interferences by laying out components in three dimensions.

Model basin. A laboratory used to predict resistance characteristics of full-size vessels, and perform other marine hydrodynamic research and design studies.

Modulus of elasticity. The stress that would double the length of a specimen if it did not break in two first. Same as **Young's modulus.**

Molded displacement. The imaginary weight of water that would be displaced by the volume within the shell plating below the water line.

Molded lines. Lines defining the geometry of a hull as a surface without thickness; structural members are related to molded lines according to standard practice (unless otherwise shown on drawings); for example, the inside surface of flush shell plating is on the molded line, also the underside of deck plating.

Mold loft. A floor space once used for laying down (laying off) the full-size lines of a ship and for making templates to lay out the hull structural components.

Moment of inertia. A measure of resistance to rotational acceleration.

Monohull. A single hull, in contradistinction to a catamaran or trimaran.

Moon pool. A large opening through the deck and bottom of a drill ship at about amidships to accommodate the major drilling operations.

Mooring. Securing a ship at a pier or elsewhere by several lines or cables so as to limit its movement.

Mooring buoy. A floating structure firmly anchored to the bottom and to which a ship may moor.

Mooring ring. A round or oval casting inserted in the bulwark plating through which the mooring lines, or hawsers, are passed.

Mooring winch. A mechanical device for controlling lines or cables used to secure the ship to a wharf.

Moorsom system. A former method for defining registered tonnage.

Mushroom. A cover permanently fitted above a ventilator located in the weather, usually round and of larger diameter than the ventilator.

Mushroom anchor. A stockless anchor resembling in shape an inverted mushroom.

Nautical mile. (See **knot.**)

Naval architect. An engineer with specific competence in designing floating craft.

Navigating bridge. The control center of a ship.

Negative stability. The condition of a floating body that will not return to its initial position if slightly deflected, but will continue to move away from that position and possibly capsize.

Net tonnage. (See **tonnage, net.**)

Network flow. A technique for relating tasks in the process of scheduling ship production or other stepwise process; the principles lend themselves to computer application.

Neutral axis. The imaginary plane within a beam or girder that is in neither tension nor compression.

Neutral helm. A condition in which a sailboat can be kept on course with the tiller kept amidships.

Neutral stability. The condition of a floating body that, if slightly deflected, will remain in the deflected position.

Node. In a vibrating structure, a place that holds still.

Numerical analysis. Mathematical methods for analyzing complex shapes in the absence of equations defining those shapes.

OBO. Abbreviation for a vessel designed to carry oil, bulk cargoes, or ore cargoes.

Offsets. A term used for the coordinates of a ship's form, deck heights, etc.

Oil stop or water stop. A detail in the construction of the intersection of members, one or both of which are oiltight, to permit continuous welding of the boundary of the tight member; usually a small semicircular cut in the edge of the "through" member near the intersection; in a lapped seam, a slot in one member, filled with weld metal.

One-compartment subdivision. A standard of subdivision of a ship by transverse bulkheads, which will result in the ship remaining afloat with any one compartment flooded, under specified conditions as to permeability of the compartment and the draft of the ship before flooding of the compartment.

Open floor. (See, **floor, open.**)

Open-water efficiency. The ratio of useful energy produced by a screw propeller to the energy put into it when operating without interference from the hull.

Outboard. Abreast or away from the centerline; outside the hull.

Outfitting. Collectively, the nonstructural components of a hull. May include hull engineering.

Overhang. That portion of a ship's bow or stern clear of the water which projects beyond the forward and after perpendiculars.

Pad eye. A fitting having one or more eyes to provide means of securing blocks, wire rope, or fiber line.

Padding. An inert gas, usually nitrogen, introduced into a cargo tank above the liquid cargo to prevent the cargo coming in contact with air.

Panel line. A production line where individual plates, framing members, webs, etc., are successively welded together to form an assembly unit which may include some items of outfit.

Panting. The pulsation in and out of the bow and stern plating as the ship alternately rises and plunges deep into the water. May also occur abreast the propellers of a multiscrew ship.

Panting frames. The frames in the forward and after portions of the hull, to prevent panting action of the shell plating.

Parabolic. Having the form of a parabola. In nontechnical terms, a curve of uniformly changing radius.

Paragraph ship. A vessel designed to measure slightly less than the gross tonnage marking the boundary for application or provisions of a convention, treaty, law or regulation. The word "paragraph" is thought to be used in the sense that tonnage boundaries are, at times, cited as provisions of a specific paragraph of a treaty, law, or regulation.

Parallel middlebody. The amidship portion of a ship within which the contour of the underwater hull form is unchanged.

Paravane. A foil, that, when towed from a vessel, will swing off to one side.

Peak. (See **afterpeak, forepeak.**)

Period of roll. The time occupied in performing one complete oscillation or roll of a ship as from port to starboard and back to port.

Permeability. The degree to which a damaged compartment can accept water.

Perpendiculars. Imaginary vertical lines drawn at the ends of the ship.

Petty officer. A seafarer with rank between a licensed officer and a rating.

Pier. A structure extending out into the water at right angles to the shore.

Pillar. Vertical member or column giving support to a deck girder, flat, or similar structure: also called **stanchion.**

Pintles. The pins or bolts that hinge the rudder to the gudgeons on the sternpost or rudderpost.

Pitch. (See **propeller pitch.**)

Pitch angle. Degrees of twist of a propeller blade relative to a plane normal to the shaft.

Pitching. The oscillatory (teeter-totter) motion of a vessel, with bow and stern moving vertically in opposite directions.

Pitch poling. The action of a boat capsizing end-over-end.

Planimeter. A mechanical device for measuring the area enclosed within a curve.

Planing hull. A boat which, when running at speed, derives lift from the pressure of water on the bottom.

Planking. Wood covering for decks, etc.

Plasma-arc cutting. A process employing an extremely high-temperature, high-velocity constricted arc between an electrode within a torch and the metal to be cut. The intense heat melts the metal, which is continuously removed by a jet-like stream of gas issuing from the torch.

Platform. (See **deck** and **flat.**)

Plimsoll mark. Informal expression for the loadline mark.

Plug weld. Where two pieces of metal overlap, they may be joined by using an arc weld to fill small holes in one of the plates, thus fusing the two plates wherever such a hole had been.

Pontoon. A hollow box, usually of steel.

Poop. A superstructure fitted at the after end of the upper deck.

Port, cargo. An opening in the side shell plating provided with a watertight cover or door and used for loading and unloading.

Porthole, portlight. (See **air port.**)

Port side. The left-hand side of a ship when looking forward. Opposite to **starboard.**

Potable. Fit for drinking. (Rhymes with notable.)

Prime mover. Device in which some source of energy is converted into mechanical power.

Prismatic coefficient. The ratio of a ship's displaced volume of water to that of a prism with a base area equal to the maximum sectional area multiplied by the vessel's length.

Propeller. A revolving screw-like device that drives the ship through the water, consisting of two or more blades; sometimes called a **screw** or **wheel.**

Propeller boss. (See **bossing.**)

Propeller disk. A plane circle encompassing the sweep of the propeller blades.

Propeller guard. A protuberant fitting that acts to protect a screw propeller from striking, or being struck by, exterior objects.

Propeller pitch. Theoretical linear distance the propeller would move ahead during one complete revolution if it were turning within a nut.

Propeller post. (See **stern frame.**)

Propeller shaft. The cylindrical member that carries rotational energy from the engine to the propeller.

Propeller tunnel. A concave shape in a bottom shell to clear a propeller, the top of which may be above the static waterline.

Proportional limit. The maximum stress to which material may be subjected and still return to its unstressed dimension when the load is removed. Also called **elastic limit.**

Propulsive coefficient. The product of propeller efficiency and hull efficiency. Used to relate effective horsepower and power delivered to the propeller.

Propulsor. Any mechanical device for imparting thrust.

Quartering sea. A series of waves approaching a vessel at about 45 deg off the bow or stern.

Quarters. Living or sleeping rooms.

Quarter wheels. Paddle wheels mounted in pairs at the stern of a vessel and capable of independent operation.

Quay. A masonry ship mooring structure usually built along the shore.

Quenching. In steelmaking, an operation consisting of heating the material to a certain temperature and holding it at that temperature to obtain desired crystalline structure, and then rapidly cooling it in a suitable medium, such as water or oil. Quenching is often followed by **tempering.**

Rabbet. A groove, depression, or offset in a member into which the end or edge of another member is fitted, generally so that the two surfaces are flush. A rabbet in the stem or stern frame would take the ends or edges of the shell plating, resulting in a flush surface.

Radius of gyration. A measure of a body's distribution of mass about an axis of rotation.

Rail. The rounded member at the upper edge of the bulwark, or the horizontal pipes or chains forming a fence-like railing fitted instead of a bulwark.

Rake. A term applied to the fore-and-aft inclination from the vertical of a mast, smokestack, stem, etc. In river, and some ocean, barges, the end portion of the hull, in which the bottom rises from the midship portion to meet the deck at the headlog.

Raked blades. Propeller blades that slope aft when viewed from the side.

Range of stability. The angle of heel to which a vessel may go before capsizing.

Rating. An unlicensed member of the crew.

Reduction gear. Mechanical device for diminishing rotational speed between prime mover and propeller.

Reefer. Colloquial abbreviation for "refrigerator" or "refrigerated."

Registered tonnage. Legally prescribed measure of a vessel's internal volume in units of 100 cubic feet. (See **tonnage, net** and **tonnage, gross.**)

Residual resistance. Inhibition to forward motion arising from energies going into creating waves and eddies. Also total resistance minus frictional resistance.

Resistance. A word of many meanings, but which to a naval architect generally pertains to the force exerted by the water on the hull in opposition to forward motion.

Resonance. A condition in which an external force is applied at periodic intervals that coincide with an object's natural period of vibration or other motion.

Resorcinol. A chemical used in the synthesis of certain resins.

Reverse frame. A bar forming the top member of an open floor, attached to the underside of the innerbottom.

Reversible-pitch propeller. A screw propeller in which the pitch of the blades can be switched from a single ahead position to a single astern position by controls within the vessel.

Ribband. A fore-and-aft wooden strip or heavy batten used to align the transverse frames and keep them in a fair line during construction. Any similar batten for fairing a ship's structure. Also, a layer of insulation on the boundaries of decks in insulated spaces. Also, steel plating or heavy planking bolted to the outboard sides of ground launching ways and projecting above the sliding surface sufficiently to keep the sliding ways in approximate alignment.

Rider plate. In singlebottom ships, a continuous flat plate attached to the top of the center vertical keel.

Rigging. Chains, wire ropes, fiber lines, and associated fittings and accessories used to support masts and booms used for handling cargo and stores and for other purposes.

Right-hand propeller. A screw propeller that provides forward thrust when turning in the clockwise direction when viewed from astern.

Rise of floor. (See **deadrise.**)

Rivet. A short metal cylinder used to pin together two or more overlapping structural members.

Roll. To impart cylindrical curvature to a plate. Also the transverse angular motion of the ship in waves. (See **period of roll.**)

Rolled shapes. Metallic components that are shaped by being squeezed between special rolls while red hot.

Roll on/roll off ship. See next entry.

RO-RO or RO/RO. Abbreviation for a vessel designed to carry vehicles, so arranged that the vehicles may be loaded and unloaded by being rolled on or off on their own and/or auxiliary wheels, via ramps fitted in the sides, bow or stern of the vessel.

Round of beam. (See **camber.**)

Rudder. A device used to steer a ship. The most common type consists of a vertical metal fin, hinged at the forward edge to the sternpost or rudderpost.

Rudder pintle. (See **pintles.**)

Rudderpost. (See **sternpost.**)

Rudderstock. A vertical shaft that connects the rudder to the steering engine.

Rudder stop. Rugged fitting on stern frame or a stout bracket on deck at each side of the quadrant, to limit the swing of the rudder to either side. A rudder angle of 35 deg is the maximum usually used at sea (45 deg on inland waterways vessels).

Run. That part of a ship's underwater body aft of the parallel middle-body or the point at which the slope of the sectional area curve is zero.

Running rigging. Sails, spars, and lines that are moved when sailing a vessel or operating cargo gear.

Sacrificial anode. Metal parts fitted to the hull of a ship to provide a transfer of ions to the cathodic part of an electrolytic coupling and so protect other parts of the ship that would otherwise waste away through electrolysis.

Sagging. Straining of the ship that tends to make the middle portion lower than the bow and stern. (See **hogging.**)

Sailboard. The equivalent of a surfboard fitted with a sail on a flexibly mounted mast that can be tilted forward or aft so as to eliminate the need for a rudder.

Samson post. (See **kingposts.**)

Scantling draft. The maximum draft at which a vessel complies with the governing strength requirements. Usually used when the scantling draft is less than the geometrical draft corresponding to the freeboard calculated according to the Load Line Convention.

Scantlings. The cross-sectional dimensions of a ship's frames, girders, plating, etc.

Scarf. A connection made between two pieces by tapering their ends so that they fit together in a joint of the same breadth and depth as the pieces connected. It is used on bar keels, stem and stern frames, and other parts.

Schilling rudder. A proprietary steering system featuring a specially shaped rudder arranged to turn at a wide angle.

Schneekluth duct. A hull appendage fitted forward of a single-screw propeller to minimize variations in the wake entering the propeller.

Schooner. A sailing vessel of whatever size fitted with two or more masts carrying fore and aft sails. When it is fitted with two masts, the forward mast may be shorter than the other.

Screen bulkhead. (See **bulkhead, screen.**)

Screw propeller. The most common marine device for converting torsional energy into thrust.

Scullery. Where the dishes, etc., are washed and dried.

Scuppers. Drains from decks to carry off accumulations of rainwater, condensation, or seawater. Scuppers are located in the gutters or waterways, on open decks, and connect to pipes usually leading overboard.

Scuttle. A small circular or oval opening fitted in decks to provide access. When used for escape and fitted with means whereby the covers can be opened quickly to permit exit, they are called quick-acting scuttles. Sometimes used to refer to an air port.

Scuttlebutt. A container for drinking water. A drinking fountain. Colloquially, rumors heard at the drinking fountain.

Sea chest. An enclosure, attached to the inside of the underwater shell and open to the sea, fitted with a portable strainer plate. A sea valve and piping connected to the sea chest pass seawater into the ship for cooling, fire, or sanitary purposes. Compressed air or steam connections may be provided to remove ice or other obstructions.

Sea-kindly. Characteristics of a vessel that remains relatively comfortable despite heavy seas.

Seam. Fore-and-aft joint of shell plating, deck, and tank top plating, or a lengthwise edge joint of any plating.

Sectional area curve. A curved line showing, at any fore and aft location, the cross-sectional area of the underwater hull form. The curve indicates how the displaced volume is distributed along the length of the ship.

Section modulus. A measure of a beam's ability to resist bending loads without failure.

Sections. A general term referring to structural bars, rolled or extruded in any cross section, such as angles, channels, bulbs, T, H, and I-bars (or beams). Sometimes called **profiles.** Also the intersections with the hull of transverse planes perpendicular to the centerline plane of the ship.

Shaft bearings. Supports along the length of a propeller shaft.

Shaft horsepower. The power delivered to the propeller shaft at the end next to the propulsion machinery.

Shaft strut. (See **strut.**)

Shaft tunnel, shaft alley. A watertight enclosure for the propeller shafting, large enough to walk in, extending aft from the engine room, to provide access and protection to the shafting in way of the holds.

Shank. In an anchor, the straight section between the chain attachment and the flukes.

Shape. A rolled bar of constant cross section such as an angle, bulb angle, channel, etc.; also to impart curvature to a plate or other member.

Shear. Parallel forces acting in opposite directions.

Sheer. The longitudinal curve of a vessel's decks in a vertical plane, the usual reference being to the ship's side; in the case of a deck having a camber, its centerline sheer may also be given in offsets. Owing to sheer, a vessel's deck height above the baseline is usually higher at the ends than amidships.

Sheer plan. The profile view in a lines drawing or lines plan. Shows shapes of buttocks.

Sheer strake. The course of shell plating at strength deck level.

Sheet. A line used to restrain the angle of a fore and aft sail.

Shell expansion. A plan showing the seams and butts, thickness, and associated welding of all shell plates.

Shell landings. Points on the frames where the edges of shell plates are located.

Shell plating. The plates forming the outer side and bottom skin of the hull.

Shell stringer. A longitudinal web frame fitted to the side shell and, usually, lending support to transverse side frames.

Shelter deck. Formerly, a term applied to a superstructure deck running continuously from stem to stern and fitted with at least one tonnage opening.

Shelter deck ship. Formerly, a ship having two or more complete decks, with scantlings and weather deck closing arrangements based on the uppermost deck being a superstructure deck and the second deck being the freeboard deck, in contrast to a full scantling ship. The purpose of the arrangement was to reduce the ship's tonnage. Under the current tonnage rules there is no longer any reason to build a shelter deck ship.

Shifting boards. Portable bulkhead members, generally constructed of wood planking and fitted fore and aft in cargo holds when carrying grain or other cargo that might shift to one side when the ship is rolling in a seaway.

Ship's service. That collection of systems outside the engine room required to operate the vessel.

Shoal water. Shallow water.

Shoe. A heavy longitudinal brace supporting the lower edge of the rudder.

Shore. A brace or prop used for support while building or repairing a ship.

Short ton. A unit of weight equal to 2000 lb.

Shot. A length of anchor chain equivalent to 15 fathoms or 90 ft.

Shroud. One of the principal members of the standing rigging, consisting of wire rope that extends from the mast head to the ship's side, affording lateral support for a mast.

Side thruster. (See **thruster.**)

Sight edge. The visible edge of shell plating as seen from outside the hull.

Simpson's multipliers. A methodical series of numbers which, when applied to an array of measurements, will provide products that are needed in applying Simpson's rule. See below.

Simpson's rule. A numerical integration technique for analyzing areas within curves without benefit of equation defining the curves.

SI units. The system of units now being used internationally is the Systeme International d'Unites (SI). Informally referred to as the metric system.

Skeg. A deep, vertical, finlike projection on the bottom of a vessel near the stern, installed to support the lower edge of the rudder, to support the propeller shaft for single-screw ships, and for the support of the vessel in dry dock: also used in pairs on barges to minimize yawing.

Skewed blades. Propeller blades that are curled back within a plane normal to the shaft line.

Skids. A skeleton framework used to hold structural assemblies above ground.

Skin tank. A tank for liquid cargo or ballast one or more of whose boundaries is the side or bottom plating of the hull.

Skylight. A framework fitted over a deck opening and having hinged covers with glass inserted for the admission of light and air to the compartment below.

Slamming. Heavy impact resulting from a vessel's bottom near the bow making sudden contact with the sea surface after having risen on a wave. Similar action results from rapid immersion of the bow in vessels with large flare.

Sliding ways. Runners riding atop the ground ways and which move with the ship during launching.

Slip. The linear distance between the pitch (or advance) and the actual distance the screw propeller moves straight ahead through the water.

Sloop. A sailboat with single mast and fore-and-aft rig, usually with a single jib plus main sail.

Smokestack. A chimney through which combustion products are led from propulsion and auxiliary machinery to the open air. Also called a **funnel.**

Snipe. A cutaway part.

Solid floor. (See **floor, solid.**)

Sounding tube. A pipe leading to the bottom of a bilge, doublebottom, deep tank, drainwell, hold, or other compartment, used to guide a sounding line, tape, or rod to determine the depth and nature of any liquid therein.

Spar deck. An anachronism used on the Great Lakes to indicate the weather or upper deck; so used because the term "main deck" is applied to the narrow plating forming the top of the usual side tanks abreast the cargo spaces on typical Great Lakes bulk carriers.

Spectacle frame. A large casting extending outboard from the main hull and furnishing support for the ends of the propeller shafts in a multiscrew ship. The shell plating (bossing) encloses the shafts and is attached at its after end to the spectacle frame. Used in place of shaft struts.

Speed-length ratio. The number found by dividing the ship's speed in knots by the square root of the length in feet. A dimensionally dependent measure of relative speed.

Spot welding. A method for lightly fastening two pieces of sheet metal. A machine presses the two pieces between small-diameter rods, between which an electric current is passed, the heat of which serves to melt and fuse the two pieces of sheet metal at that point.

Spray strip. A narrow longitudinal appendage fitted to the forebody of a high-speed boat to deflect waves and suppress spray.

Spring bearing. Bearings to support the line shafting.

Springing. A vibration of the complete vessel induced by wave forces in conjunction with the ship's elastic properties. More pronounced in ships having a high length-to-depth ratio.

Spring line. A mooring rope oriented at a small angle to the ship's centerline.

Square propeller. A screw propeller in which the pitch equals the diameter.

Squatting. The increase in trim by the stern assumed by a ship when underway over that existing when at rest.

Stability. The tendency of a ship to remain upright or the ability to return to the normal upright position when heeled by the action of waves, wind, etc.

Staging. Horizontal surfaces giving shipbuilding personnel access to the work, either inside or outside the hull; may be supported by temporary members attached to subassemblies or by wood or metal uprights, usually portable.

Stanchion. Vertical column supporting decks, flats, girders, etc.; also called a pillar. Rail stanchions are vertical metal columns on which fence-like rails are mounted. (See **rail.**)

Standard section. Any rolled structural member in a form readily available from a mill.

Standard series. Families of methodically varied hull forms subjected to model tests, with results used to predict wave-making resistance of proposed vessels. Also used for propellers.

Standing rigging. Fixed rigging supporting the masts such as shrouds and stays. Does not include running rigging such as boom topping lifts, vangs, and cargo falls.

Starboard side. The right-hand side of a ship when looking forward. Opposite to **port.**

Starting air. Compressed air used to start a diesel engine.

Static stability. The tendency for a body to remain upright when floating in calm water with no external disturbing influences.

Station. A fixed position along the length of a vessel. The transverse shapes of the stations as they intersect the molded hull form are shown on the body plan, a part of the lines drawing.

Stays. Fixed wire ropes leading forward from aloft on a mast to the deck to prevent the mast from bending aft. Backstays lead from aloft to the deck edge well abaft the position of the mast. Preventer stays lead to any point on the deck to provide additional mast support when handling heavy loads with boom tackle.

Steady bearing. (See **spring bearing.**)

Stealer. A single wide plate that is butt-connected to two narrow plates, usually near the ends of a ship.

Steaming profile. A systematic outline of the power requirements during a typical voyage.

Steam reciprocating engine. Another name for an old-fashioned steam engine: a mechanical device using pistons acting within cylinders to convert heat energy (in steam) to rotating mechanical energy.

Steam turbine. A prime mover that provides mechanical power using steam from a boiler to drive internal windmill-like blades.

Steering gear. A term applied to the steering wheels, leads, steering engine, and fittings by which the rudder is turned. Usually applied to the steering engine.

Steering nozzle. A turnable shroud fitted around a screw propeller to provide steering ability as well as thrust enhancement.

Stem. The bow frame forming the apex of the intersection of the forward sides of a ship. It is rigidly connected at its lower end to the keel and may be a heavy flat bar or of rounded plate construction.

Step, mast. (See **mast step.**)

Stern. After end of ship.

Stern, clearwater. A stern with a "shoeless" stern frame.

Stern, cruiser. A spoon- or canoe-shaped stern used on many merchant ships, designed to give maximum immersed length.

Stern frame. Large casting, forging or weldment attached to after end of the keel. Incorporates the rudder gudgeons and in single-screw ships includes the propeller post.

Stern gland. A fitting encompassing the tailshaft where it passes through the afterpeak bulkhead, allowing soft material to be squeezed against the shaft to prevent entry of seawater into the ship.

Sternpost. Sometimes, the vertical part of the stern frame to which the rudder is attached.

Stern, transom. A square-ended stern used to provide additional hull volume and deck space aft, or to decrease resistance in some high-speed ships.

Stern tube. The watertight tube enclosing and supporting the tailshaft. It consists of a cast-iron or cast-steel cylinder fitted with bearing surface within which the tailshaft rotates.

Stiff, stiffness. A vessel is said to be stiff if it has an abnormally large metacentric height. Such a ship may have a short period of roll and therefore will roll uncomfortably. The opposite of **tender.**

Stiffener. An angle, T-bar, channel, built-up section, etc., used to stiffen plating of a bulkhead, etc.

Stock anchor. An anchor fitted at the upper end with a stout cross member running at 90 degrees to the line of the flukes.

Stopwater. (See **oil stop.**)

Stow. To put away. To stow cargo in a hold.

Stowage factor. The volume of a given type of cargo per unit of its weight.

Strain. The deformation resulting from a stress, measured by the ratio of the change to the total value of the dimension in which the change occurred.

Strake. A course, or row, of shell, deck, bulkhead, or other plating.

Stream anchor. An anchor carried at the stern.

Strength deck. The deck that is designed as the uppermost part of the main hull longitudinal strength girder. The bottom shell plating forms the lowermost part of this girder.

Stress. The force per unit section area producing deformation in a body.

Stringer. A term applied to a fore-and-aft girder running along the side of a ship at the shell and also to the outboard strake of plating on any deck. Also the side pieces of a ladder or staircase into which the treads and risers are fastened.

Stringer plate. (See **deck stringer.**)

Strongback. A piece of plate or a special tool used to align the edges of plates to be welded together. Also a steel bar such as a channel, one or more of which may be used to secure a door in closed position, in addition to the dogs around its edge.

Strut. Outboard column-like support or V-arranged supports for the propeller shaft, used instead of bossings on some ships with more than one propeller. Also, any short structural member intended to carry compressive or tensile loads.

Stuelcken gear. Heavy lift cargo gear consisting of a pair of freestanding masts, flared, with a boom hinged between them on the centerline capable of use either abaft or forward of the masts.

Subdivision. The technology of locating watertight bulkheads along the length of a ship so as to confine flooding in case of damage to the shell. May involve decks as well.

Submerged-arc welding. A fusion welding process in which a machine feeds an electrode and a flux in granular form as the machine moves over the joint; the flux covers the arc, submerging it, and is melted to a brittle slag covering the deposited metal.

Summer load line. The basic maximum draft established by the freeboard rules.

Supercavitating propeller. A kind of screw propeller so shaped as to create a steady cavitation space and prevent cavitation bubbles from collapsing on the blades.

Superconducting. Electrical systems employing extremely cold elements in order to reduce electrical resistance and so increase efficiency.

Superstructure. A decked-over structure above the upper deck, the outboard sides of which are formed by the shell plating as distinguished from a deckhouse that does not extend outboard to the ship's sides. (See **shelter deck,** also **poop, bridge** and **forecastle.**)

Surging. Condition of a vessel experiencing variable ahead speed.

Swash bulkhead, swash plate. Longitudinal or transverse nontight bulkheads fitted in a tank to decrease the swashing action of the liquid contents as a ship rolls and pitches at sea. Their function is greatest when the tanks are partially filled. Without them the unrestricted action of the liquid against the sides of the tank might be severe. A plate serving this purpose but not extending to the bottom of the tank is called a **swash plate.**

Swaying. Condition of a vessel experiencing lateral translational motions.

Systems analysis. A methodical design procedure aimed at decisions that will maximize the object's value in the face of the constraints. Considers operating procedures as well as hardware design.

Tack, tacking. The action by which a sailing vessel can work its way against an adverse wind by sailing first at one angle off the wind, and then turning into the wind and sailing off at the opposite angle on a zigzag course.

Tactical diameter. The transverse distance effected by a vessel after start of the hard-over turning maneuver. Measured when the heading has changed 180 deg from original course.

Tailshaft. The aftermost section of the propulsion shafting, in the stern tube in single-screw ships and in the struts of multiple-screw ships, to which the propeller is fitted.

Tank, ballast. (See **ballast tank.**)

Tanker. A cargo ship intended to carry liquid cargo in bulk.

Tank, peak. (See **afterpeak, forepeak.**)

Tank, settling. Fuel oil tanks used for separating entrained water from the oil. The oil is allowed to stand for a few hours until the water has settled to the bottom, when the latter is drained or pumped off.

Tanktop. (See **innerbottom.**)

Tank, trimming. A tank located near the ends of a ship. Seawater (or fuel oil) is carried in such tanks as necessary to change trim.

Tank, wing. Tanks located well outboard adjacent to the side shell plating, often consisting of a continuation of the doublebottom up the sides to a deck or flat.

T-bar. A stiffening member shaped, in section, like the letter T.

Telegraph. An apparatus, either electrical or mechanical, for transmitting orders, as from a ship's bridge to the engine room, steering gear room, or elsewhere about the ship.

Telemotor. A device for operating the control valves of the steering engine from the pilothouse, either by fluid pressure or by electricity.

Tempering. After quenching, the material is reheated to a predetermined temperature below the critical range and then cooled. In steelmaking this is done to relieve stresses set up by quenching and to restore ductility.

Template. Full-size patterns (often paper or wood) to be placed on materials to indicate the plate edges, etc.; also to indicate the curvature to which frames, for example, are to be bent.

Tender. A vessel is said to be tender if it has an abnormally small metacentric height. Such a ship may have a long period of roll but may list excessively in a strong wind and may be dangerous if a hold is flooded. The opposite of **stiff.** Also a small general utility boat carried aboard ship or towed astern.

Tensile. Pertaining to a condition of stretching.

Teredo. A worm that inhabits and devours wood immersed in salt water.

Test head. The head or height of the column of water that will give a prescribed pressure on the vertical or horizontal sides of a compartment or tank in order to test its tightness, or strength.

Three-watch system. A method for managing a ship in which three groups of seafarers share the duties, each working eight hours per day.

Thrust deduction. A restraining force arising from negative pressure on the stern caused by propeller action.

Thruster. A mechanical device for providing (usually) a transverse force in order to help turn a ship.

Thrust recess. A small compartment at the after end of the main engine room at forward end of shaft tunnel, designed to contain and give access to the thrust shaft and thrust bearing.

Tie plate. A fore-and-aft course of plating attached to deck beams under a wood deck for strength purposes.

Tiller. An arm, attached to rudderstock, which turns the rudder.

Ton. A unit of weight, usually a long ton of 2240 pounds, or a metric ton of about 2205 pounds.

Tonne. A metric ton.

Tonnage, gross. An approximate measure of a vessel's total volume. Under vessel measurement rules of various nations, the Panama Canal and the Suez Canal, a measure of the internal volume of spaces within a vessel in which 1 ton is equivalent to 2.83 cubic meters or 100 cu ft. Under the International Convention of Tonnage Measurement of Ships (ICTM) 1969, a standardized numerical value that is a logarithmic function of spaces within a vessel. Gross tonnage according to ICTM is $GT = K_1 V$ in which V is the total molded volume of all enclosed spaces of the ship in cubic meters and K_1 is $0.2 + 0.02 \log_{10} V$.

Tonnage, net. An approximation to a vessel's money-earning volume. Net tonnage according to canal rules is derived from gross tonnage by deducting an allowance for the propelling machinery space and certain other spaces. Net tonnage according to ICTM is a logarithmic function of the volume of cargo space, the draft-to-depth ratio, the number of passengers to be accommodated, and the gross tonnage.

Tonnage opening. A device formerly used to circumvent the tonnage measurement rules, consisting of a closure just short of meeting the definition of a watertight closing appliance, but nevertheless accepted in the United States as a weathertight closing appliance for purposes of the Load Line regulations.

Topping lift. A wire rope or tackle extending from the head of a boom to a mast, or to an elevated part of the ship's structure, for the purpose of supporting the weight of the boom and its loads, and permitting the boom to be raised or lowered.

Topside tank. A tank, triangular in section, usually used for ballast in bulk carriers, formed by the deck outboard of the hatches, the upper shell plating, and a longitudinal bulkhead sloping around 45 deg.

Toughness. The ability of a material to withstand shock.

Tramp ship. A cargo ship that has no set trade route or schedule.

Transfer. At any point in the hard-over turning maneuver, the lateral distance moved off the original course.

Transformation temperature. The temperature above which the ferrite form of iron in shipbuilding and other steel is transformed to the austenite form and below which the ferrite form recurs. The microstructure of the steel is changed upon passing through this temperature.

Transmission system. The collection of devices used to connect the prime mover to the propelling device.

Transom beam. The aftermost transverse deck beam. Abaft and connected to it are the cant beams. Connects at its ends to the transom frame.

Transom frame. The aftermost transverse side frame, abaft which are the cant frames.

Transom stern. (See **stern, transom.**)

Transverse. A deep member supporting longitudinal frames of bottom or side shell or longitudinal deck beams. At right angles to the fore-and-aft centerline.

Transverse framing. A system of hull stiffening in which the stiffening members run across the deck or decks, down the sides, and across the bottom.

Transverse stability. That branch of stability having to do with the tendency for a vessel to resist heeling and capsize.

Trawler. A vessel that catches fish by dragging a net along the floor of the ocean.

Trim. The difference between the draft forward and the draft aft. If the draft forward is the greater, the vessel is said to "trim by the head." If the draft aft is the greater, it is "trimming by the stern." To trim a ship is to adjust the location of cargo, fuel, etc., so as to result in the desired drafts forward and aft.

Trimaran. A triple-hulled boat, the center hull usually being larger than the others.

Tripping bracket. Flat bars or plates fitted at various points on deck girders, stiffeners, or beams as reinforcements to prevent their free flanges from distorting under compression.

Troichoid. The curve traced out by a point on the radius of a circle as the circle is rolled along a straight line.

Trunk. A vertical or inclined space or passage formed by bulkheads or casings, extending one or more deck heights, around openings in the decks, through which access can be obtained and cargo, stores, etc., handled, or ventilation provided without disturbing or interfering with the contents or arrangements of the adjoining spaces.

Tumblehome. Inboard slope of a ship's side, usually above the designed waterline.

Tumble shore. A strut, perpendicular to the plane of the launch ways, between the bottom of a vessel about to be launched and the ground or building slip deck. Used to keep some of ship weight off the launching grease until sliding starts. Motion of the vessel causes the shore to fall or tumble to the ground. Typically, one or more tumble shores might be installed under the outshore end of the keel of a ship to be end launched.

Tunnel, propeller. (See **propeller tunnel.**)

Tunnel, shaft. (See **shaft tunnel.**)

Tunnel stern. The after end of a vessel shaped with one or more propeller tunnels.

Turbine. Usually **steam turbine:** A high-rpm kind of propulsion device deriving energy from steam impinging on radial blades. (See also **gas turbine.**)

'Tween decks. The space between any two adjacent decks.

Twist lock. A mechanical locking device at the corner of a cargo container. Rotating the movable part about a vertical axis locks the device and container to a mating component on an above-deck stowage or another container. Reversing the motion effects unlocking.

Two-stroke engine. In reciprocating prime movers, one in which the energy is applied once during every complete revolution.

Two-watch system. A ship manning method in which two groups of seafarers take turns operating the ship.

Universal chock. A deck-edge fitting through which a mooring line may be fed and which allows easy passage of the line in any direction.

Uptake. A casing connecting a boiler or gas turbine combustion product outlet with the base of the inner casing of the smokestack.

Vang. Wire rope or tackle secured to the end of a cargo boom, the lower end being secured to the deck, top of bulwark, or to a special post at the ship's side. Used to swing the boom and hold it in a desired position.

V-bottom. Lowermost part of a hull with extreme deadrise.

Ventilator, cowl. (See **cowl.**)

Viscosity. That property of fluids resulting in resistance to flow.

VLCC tanker. A very large crude oil carrier, over about 150 000 DWT.

Voith Schneider wheel. A proprietary kind of cycloidal propeller.

Wake. The body of water that tends to follow a ship. It is set in motion by friction with the hull.

Waterline. The line of the water's edge when the ship is afloat; technically, the intersection of any horizontal plane with the molded form.

Waterlines plan. A part of the lines drawing on which are shown the shapes of the waterplanes.

Waterplane. The area encompassed by a fixed vertical location in the hull form.

Waterplane coefficient. The ratio of the area of a waterplane divided by the product of the ship's length and beam.

Waterway. A narrow gutter along the edge of a deck for drainage.

Wave-making resistance. Inhibition to forward motion through the water arising from the energy required to generate waves created by the hull.

Ways. (See **ground ways** and **sliding ways.**)

Weather deck. (See **deck, weather.**)

Weather helm. A condition in which a sailboat can be kept on course only by turning the tiller somewhat upwind.

Weather routing. Selecting a course with due regard to predicted climatic conditions.

Weathertight. Ability to shed casual water, such as rain, that has no pressure.

Web. In a structural member, that part that resists shear loads and serves to separate the flanges. The term is also used to mean a web frame.

Web frame. A built-up frame to provide extra strength, usually consisting of a web plate, flanged or otherwise stiffened on its edge.

Weedless propeller. A screw propeller with extremely skewed blades.

Well. Space in the double bottom of a ship to which bilge water drains so that it may be pumped overboard; also space between partial superstructures.

Wetted surface. The area of the immersed shell plating, plus that of the appendages.

Wharf. A general term for a place to moor a ship. It may be either a pier or a quay.

Wheel. Often used informally to mean propeller. Also, the steering wheel.

Wheelhouse. Another name for the control center of the ship.

Whip. In cargo handling, the wire leading to the hook by which the draft of cargo is being hoisted.

Wide flange. A structural member shaped much like an I-beam, but with much wider flanges.

Wildcat. A special type of cog-like windlass drum whose faces are formed to fit the links of the anchor chain. The rotating wildcat causes the chain to be slacked off when lowering the anchor, or hauled in when raising it.

Winch. A machine, usually steam or electric, used primarily for hoisting and lowering cargo but also for other purposes.

Windlass. The machine used to hoist and lower anchors.

Wing keel, winglets. A keel fitted with twin fixed appendages at its bottom, often in an inverted V-shape.

Yacht. A vessel of any size, intended for racing, cruising, or other recreational activity.

Yawing. Weaving motion of a vessel to port and starboard off course.

Yawl. A boat with fore-and-aft sailing rig with two masts, the forward one being considerably larger than the other.

Yield stress. The maximum stress to which material may be subjected and still return to its original dimension when the load is removed. It differs slightly from the **proportional limit** in that yield stress may be reached at unit loads slightly higher than the range wherein stress and strain are directly proportional.

Young's modulus. (See **modulus of elasticity.**)

INDEX